THE APPLIED ARTIFICIAL INTELLIGENCE WORKSHOP

Start working with AI today, to build games, design decision trees, and train your own machine learning models

Anthony So, William So, and Zsolt Nagy

THE APPLIED ARTIFICIAL INTELLIGENCE WORKSHOP

Authors: Anthony So, William So, and Zsolt Nagy

Reviewers: Swanand Bagve, John Wesley Doyle, Ashish Pratik Patil, Shantanu Shivrup Pathak, and Subhranil Roy

Managing Editor: Adrian Cardoza

Acquisitions Editors: Manuraj Nair, Sneha Shinde, Anindya Sil, and Karan Wadekar

Production Editor: Salma Patel

Editorial Board: Megan Carlisle, Samuel Christa, Mahesh Dhyani, Heather Gopsill, Manasa Kumar, Alex Mazonowicz, Monesh Mirpuri, Bridget Neale, Dominic Pereira, Shiny Poojary, Abhishek Rane, Brendan Rodrigues, Erol Staveley, Ankita Thakur, Nitesh Thakur, and Jonathan Wray

First published: July 2020

Production reference: 2230221

ISBN: 978-1-80020-581-9

Published by Packt Publishing Ltd.

Livery Place, 35 Livery Street

Birmingham B3 2PB, UK

WHY LEARN WITH A PACKT WORKSHOP?

LEARN BY DOING

Packt Workshops are built around the idea that the best way to learn something new is by getting hands-on experience. We know that learning a language or technology isn't just an academic pursuit. It's a journey towards the effective use of a new tool—whether that's to kickstart your career, automate repetitive tasks, or just build some cool stuff.

That's why Workshops are designed to get you writing code from the very beginning. You'll start fairly small—learning how to implement some basic functionality—but once you've completed that, you'll have the confidence and understanding to move onto something slightly more advanced.

As you work through each chapter, you'll build your understanding in a coherent, logical way, adding new skills to your toolkit and working on increasingly complex and challenging problems.

CONTEXT IS KEY

All new concepts are introduced in the context of realistic use-cases, and then demonstrated practically with guided exercises. At the end of each chapter, you'll find an activity that challenges you to draw together what you've learned and apply your new skills to solve a problem or build something new.

We believe this is the most effective way of building your understanding and confidence. Experiencing real applications of the code will help you get used to the syntax and see how the tools and techniques are applied in real projects.

BUILD REAL-WORLD UNDERSTANDING

Of course, you do need some theory. But unlike many tutorials, which force you to wade through pages and pages of dry technical explanations and assume too much prior knowledge, Workshops only tell you what you actually need to know to be able to get started making things. Explanations are clear, simple, and to-the-point. So you don't need to worry about how everything works under the hood; you can just get on and use it.

Written by industry professionals, you'll see how concepts are relevant to real-world work, helping to get you beyond "Hello, world!" and build relevant, productive skills. Whether you're studying web development, data science, or a core programming language, you'll start to think like a problem solver and build your understanding and confidence through contextual, targeted practice.

ENJOY THE JOURNEY

Learning something new is a journey from where you are now to where you want to be, and this Workshop is just a vehicle to get you there. We hope that you find it to be a productive and enjoyable learning experience.

Packt has a wide range of different Workshops available, covering the following topic areas:

- Programming languages
- Web development
- Data science, machine learning, and artificial intelligence
- Containers

Once you've worked your way through this Workshop, why not continue your journey with another? You can find the full range online at http://packt.live/2MNkuyl.

If you could leave us a review while you're there, that would be great. We value all feedback. It helps us to continually improve and make better books for our readers, and also helps prospective customers make an informed decision about their purchase.

Thank you,
The Packt Workshop Team

Table of Contents

Chapter 6: Neural Networks and Deep Learning　　267

PREFACE

ABOUT THE BOOK

You already know that **Artificial Intelligence (AI)** and **Machine Learning (ML)** are present in many of the tools you use in your daily routine. But do you want to be able to create your own AI and ML models and develop your skills in these domains to kickstart your AI career?

The Applied Artificial Intelligence Workshop gets you started with applying AI with the help of practical exercises and useful examples, all put together cleverly to help you gain the skills to transform your career.

The book begins by teaching you how to predict outcomes using regression. You'll then learn how to classify data using techniques such as **K-Nearest Nneighbor (KNN)** and **Support Vector Machine (SVM)** classifiers. As you progress, you'll explore various decision trees by learning how to build a reliable decision tree model that can help your company find cars that clients are likely to buy. The final chapters will introduce you to deep learning and neural networks. Through various activities, such as predicting stock prices and recognizing handwritten digits, you'll learn how to train and implement **Convolutional Neural Networks (CNNs)** and **Recurrent Neural Networks (RNNs)**.

By the end of this applied AI book, you'll have learned how to predict outcomes and train neural networks, and be able to use various techniques to develop AI and ML models.

AUDIENCE

The Applied Artificial Intelligence Workshop is designed for software developers and data scientists who want to enrich their projects with machine learning. Although you do not need any prior experience in AI, it is recommended that you have knowledge of high-school-level mathematics and at least one programming language, preferably Python. While this is a beginner's book, experienced students and programmers can also improve their Python programming skills by focusing on the practical applications featured in this AI book.

ABOUT THE CHAPTERS

Chapter 1, Introduction to Artificial Intelligence, introduces AI. You will also be implementing your first AI through a simple tic-tac-toe game where you will be teaching the program how to win against a human player.

Chapter 2, An Introduction to Regression, introduces regression. You will come across various techniques, such as linear regression, with one and multiple variables, along with polynomial and support vector regression.

Chapter 3, An Introduction to Classification, introduces classification. Here, you will be implementing various techniques, including k-nearest neighbors and support vector machines.

Chapter 4, An Introduction to Decision Trees, introduces decision trees and random forest classifiers.

Chapter 5, Artificial Intelligence: Clustering, really starts getting you thinking in new ways with your first unsupervised models. You will be introduced to the fundamentals of clustering and will implement flat clustering with the k-means algorithm and hierarchical clustering with the mean shift algorithm.

Chapter 6, Neural Networks and Deep Learning, introduces TensorFlow, Convolutional Neural Networks (CNNs), and Recurrent Neural Networks (RNNs). You will also be implementing an image classification program using neural networks and deep learning.

CONVENTIONS

Code words in text, folder names, filenames, file extensions, pathnames, and user input are shown as follows: "Please note that this function is in the **tensorflow** namespace, which is not referred to by default."

A block of code is set as follows:

```
features_train = features_train / 255.0
features_test = features_test / 255.0
```

New terms and important words are shown like this: "**Mean-shift** is an example of hierarchical clustering, where the clustering algorithm determines the number of clusters."

CODE PRESENTATION

Lines of code that span multiple lines are split using a backslash (\ ). When the code is executed, Python will ignore the backslash, and treat the code on the next line as a direct continuation of the current line.

For example:

```
history = model.fit(X, y, epochs=100, batch_size=5, verbose=1, \
                    validation_split=0.2, shuffle=False)
```

Comments are added into code to help explain specific bits of logic. Single-line comments are denoted using the # symbol, as follows:

```
# Print the sizes of the dataset
print("Number of Examples in the Dataset = ", X.shape[0])
print("Number of Features for each example = ", X.shape[1])
```

Multi-line comments are enclosed by triple quotes, as shown below:

```
"""
Define a seed for the random number generator to ensure the
result will be reproducible
"""
seed = 1
np.random.seed(seed)
random.set_seed(seed)
```

SETTING UP YOUR ENVIRONMENT

Before we explore the book in detail, we need to set up specific software and tools. In the following section, we shall see how to do that.

INSTALLING JUPYTER ON YOUR SYSTEM

To install Jupyter on Windows, MacOS, and Linux, perform the following steps:

1. Head to https://www.anaconda.com/distribution/ to install the Anaconda Navigator, which is an interface through which you can access your local Jupyter notebook.

2. Now, based on your operating system (Windows, MacOS, or Linux), you need to download the Anaconda Installer. Have a look at the following figure where we have downloaded the Anaconda files for Windows:

Getting started with Anaconda

Expedite your data science journey with easy access to training materials, documentation, and community resources including Anaconda Cloud

DOCUMENTATION
Review documentation for Anaconda
Individual Edition.

STARTER VIDEO
Watch a short video to get started using
Individual Edition.

TRAINING
Sign up for online DataCamp classes to learn
Python for data science.

SUPPORT
Have a question or need to submit a pull
request? Visit our Github page.

Anaconda Installers

Windows	MacOS	Linux
Python 3.7	Python 3.7	Python 3.7
64-Bit Graphical Installer (466 MB)	64-Bit Graphical Installer (442 MB)	64-Bit (x86) Installer (522 MB)
32-Bit Graphical Installer (423 MB)	64-Bit Command Line Installer (430 MB)	64-Bit (Power8 and Power9) Installer (276 MB)
Python 2.7	Python 2.7	
64-Bit Graphical Installer (413 MB)	64-Bit Graphical Installer (637 MB)	Python 2.7
32-Bit Graphical Installer (356 MB)	64-Bit Command Line Installer (409 MB)	64-Bit (x86) Installer (477 MB)
		64-Bit (Power8 and Power9) Installer (295 MB)

Figure 0.1: The Anaconda home screen

LAUNCHING THE JUPYTER NOTEBOOK

To launch the Jupyter Notebook from the Anaconda Navigator, you need to perform the following steps:

1. Once you install the Anaconda Navigator, you will see the screen shown in *Figure 0.2*:

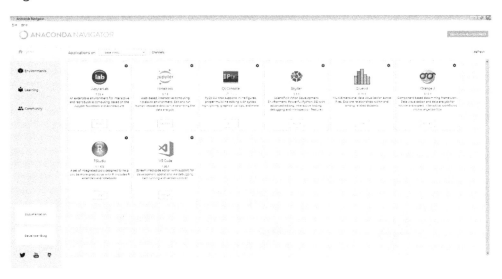

Figure 0.2: Anaconda installation screen

2. Now, click on **Launch** under the Jupyter Notebook option and launch the notebook on your local system:

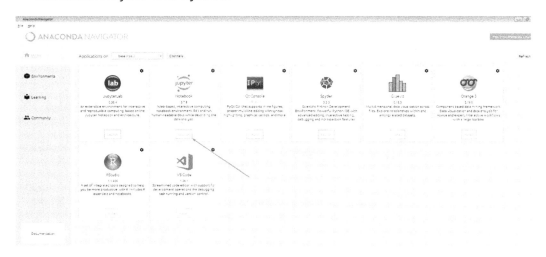

Figure 0.3: Jupyter Notebook launch option

You have successfully installed Jupyter Notebook on your system.

INSTALLING LIBRARIES

`pip` comes pre-installed with Anaconda. Once Anaconda is installed on your machine, all the required libraries can be installed using `pip`, for example, `pip install numpy`. Alternatively, you can install all the required libraries using `pip install -r requirements.txt`. You can find the **requirements.txt** file at https://packt.live/3erXq0B.

The exercises and activities will be executed in Jupyter Notebooks. Jupyter is a Python library and can be installed in the same way as the other Python libraries – that is, with `pip install jupyter`, but fortunately, it comes pre-installed with Anaconda. To open a notebook, simply run the command `jupyter notebook` in the Terminal or Command Prompt.

A FEW IMPORTANT PACKAGES

Some of the exercises in this chapter require the following packages:

- EasyAI
- Quandl
- TensorFlow 2.1.0

Install them by following this guide. On Windows, open up Command Prompt. On macOS or Linux, open up Terminal.

To install **easyAI** and **Quandl**, type the following command:

```
pip install easyAI==1.0.0.4 Quandl==3.5.0 tensorflow==2.1.0
```

ACCESSING THE CODE FILES

You can find the complete code files of this book at https://packt.live/31biHYK. You can also run many activities and exercises directly in your web browser by using the interactive lab environment at https://packt.live/2Vbev7E.

We've tried to support interactive versions of all activities and exercises, but we recommend a local installation as well for instances where this support isn't available.

If you have any issues or questions about installation, please email us at workshops@packt.com.

1

INTRODUCTION TO ARTIFICIAL INTELLIGENCE

OVERVIEW

In this chapter, you will be introduced to the fundamentals of Artificial Intelligence (AI), which are the foundations of various fields of AI. You will also come across different algorithms, including MinMax and A*, through simple coding exercises using the Python programming language. You will also be implementing your first AI through a simple tic-tac-toe game where you will be teaching the program how to win against a human player. By the end of this chapter, you will learn how to use popular Python libraries to develop intelligent AI-driven programs.

INTRODUCTION

Before discussing the different AI techniques and algorithms, we will look at the fundamentals of AI and machine learning and go through a few basic definitions. Real-world examples will be used to present the basic concepts of AI in an easy-to-digest way.

AI attempts to replicate human intelligence using hardware and software solutions. It is based on reverse engineering. For example, artificial neural networks are modeled after the way the human brain works. Beyond neural networks, there are many other models in neuroscience that can be used to solve real-world problems using AI. Companies that are known to be using AI in their fields include Google, with Google Translate, Apple, with Face ID, Amazon, with its Alexa products, and even Uber and Tesla, who are still working on building self-driving cars.

On the other hand, machine learning is a term that is often confused with AI. It originates from the 1950s and was first defined by Arthur Lee Samuel in 1959.

In his book called *Machine Learning*, Tom Mitchell proposed a simple definition of it: "*The field of machine learning is concerned with the question of how to construct computer programs that automatically improve with experience.*"

We can understand this as machine learning being the field where the goal is to build a computer program capable of learning patterns from data and improve its learning ability with more data.

He also proposed a more formal definition, which is that a computer program is said to learn from experience, **E**, with respect to a task, **T**, and a performance measure, **P**, if its performance on **T**, as measured by **P**, improves with experience, **E**. This can be translated as what a computer program requires in order for it to be learning. We can see **E** (the experience) as the data that needs to be fed to the machine, **T**, as the type of decision that the machine needs to perform, and **P** as the measure of its performance.

From these two definitions, we can conclude that machine learning is one way to achieve AI. However, you can have AI without machine learning. For instance, if you hardcode rules and decision trees, or you apply search techniques, you can create an AI agent, even though your approach has little to do with machine learning.

AI and machine learning have helped the scientific community harness the explosion of big data with more and more data being created every second. With AI and machine learning, scientists can extract information that human eyes cannot process fast enough on these huge datasets.

Now that we have been introduced to AI and machine learning, let's focus on AI.

HOW DOES AI SOLVE PROBLEMS?

AI automates human intelligence based on the way a human brain processes information.

Whenever we solve a problem or interact with people, we go through a process. By doing this, we limit the scope of a problem or interaction. This process can often be modeled and automated in AI.

AI makes computers appear to think like humans.

Sometimes, it feels like the AI knows what we need. Just think about the personalized coupons you receive after shopping online. AI knows which product we will most likely be purchasing. Machines are able to learn your preferences through the implementation of different techniques and models, which we will look at later in this book.

AI is performed by computers that are executing low-level instructions.

Even though a solution may appear to be intelligent, we write code, just like with any other software solution in AI. Even if we are simulating neurons, simple machine code and computer hardware executes the *thinking* process.

Most AI applications have one primary objective. When we interact with an AI application, it seems human-like because it can restrict a problem domain to a primary objective. Therefore, the process whereby the AI reaches the objective can be broken down into smaller and simpler low-level instructions.

AI may stimulate human senses and thinking processes for specialized fields.

You must be able to simulate human senses and thoughts, and sometimes trick AI into believing that we are interacting with another human. In some special cases, we can even enhance our own senses.

Similarly, when we interact with a chatbot, for instance, we expect the bot to understand us. We expect the chatbot or even a voice recognition system to provide a computer-human interface that fulfills our expectations. In order to meet these expectations, computers need to emulate human thought processes.

DIVERSITY OF DISCIPLINES IN AI

A self-driving car that cannot sense other cars driving on the same highway would be incredibly dangerous. The AI agent needs to process and sense what is around it in order to drive the car. However, this is not enough since, without understanding the physics of moving objects, driving the car in a normal environment would be an almost impossible, not to mention deadly, task.

In order to create a usable AI solution, different disciplines are involved, such as the following:

- **Robotics**: To move objects in space

- **Algorithm theory**: To construct efficient algorithms

- **Statistics**: To derive useful results, predict the future, and analyze the past

- **Psychology**: To model how the human brain works

- **Software engineering**: To create maintainable solutions that endure the test of time

- **Computer science or computer programming**: To implement our software solutions in practice

- **Mathematics**: To perform complex mathematical operations

- **Control theory**: To create feed-forward and feedback systems

- **Information theory**: To represent, encode, decode, and compress information

- **Graph theory**: To model and optimize different points in space and to represent hierarchies

- **Physics**: To model the real world

- **Computer graphics and image processing**: To display and process images and movies

In this book, we will cover a few of these disciplines, including algorithm theory, statistics, computer science, mathematics, and image processing.

FIELDS AND APPLICATIONS OF AI

Now that we have been introduced to AI, let's move on and see its application in real life.

SIMULATION OF HUMAN BEHAVIOR

Humans have five basic senses that can be divided into visual (seeing), auditory (listening), kinesthetic (moving), olfactory (smelling), and gustatory (tasting). However, for the purposes of understanding how to create intelligent machines, we can separate these disciplines as follows:

- Listening and speaking

- Understanding language

- Remembering things

- Thinking

- Seeing

- Moving

A few of these are out of scope for us because the purpose of this chapter is to understand the fundamentals. In order to move a robot arm, for instance, we would have to study complex university-level math to understand what's going on, but we will only be sticking to the practical aspects in this book:

- **Listening and speaking**: Using a speech recognition system, AI can collect information from a user. Using speech synthesis, it can turn internal data into understandable sounds. Speech recognition and speech synthesis techniques deal with the recognition and construction of human sounds that are emitted or that humans can understand.

  For instance, imagine you are on a trip to a country where you don't speak the local language. You can speak into the microphone of your phone, expect it to *understand* what you say, and then translate it into the other language. The same can happen in reverse with the locals speaking and AI translating the sounds into a language you understand. Speech recognition and speech synthesis make this possible.

> **NOTE**
>
> An example of speech synthesis is Google Translate. You can navigate to https://translate.google.com/ and make the translator speak words in a non-English language by clicking the loudspeaker button below the translated word.

- **Understanding language**: We can understand natural language by processing it. This field is called natural language processing, or NLP.

  When it comes to NLP, we tend to learn languages based on statistical learning by learning the statistical relationship between syllables.

- **Remembering things**: We need to represent things we know about the world. This is where creating knowledge bases and hierarchical representations called **ontologies** comes into play. Ontologies categorize things and ideas in our world and contain relations between these categories.

- **Thinking**: Our AI system has to be an expert in a certain domain by using an expert system. An expert system can be based on mathematical logic in a deterministic way, as well as in a fuzzy, non-deterministic way.

  The knowledge base of an expert system is represented using different techniques. As the problem domain grows, we create hierarchical ontologies.

  We can replicate this structure by modeling the network on the building blocks of the brain. These building blocks are called neurons, and the network itself is called a neural network.

- **Seeing**: We have to interact with the real world through our senses. We have only touched upon auditory senses so far, in regard to speech recognition and synthesis. What if we had to see things? If that was the case, we would have to create computer vision techniques to learn about our environment. After all, recognizing faces is useful, and most humans are experts at that.

  Computer vision depends on image processing. Although image processing is not directly an AI discipline, it is a required discipline for AI.

- **Moving**: Moving and touching are natural to us humans, but they are very complex tasks for computers. Moving is handled by robotics. This is a very math-heavy topic.

  Robotics is based on control theory, where you create a feedback loop and control the movement of your object based on the feedback gathered. Control theory has applications in other fields that have absolutely nothing to do with moving objects in space. This is because the feedback loops that are required are similar to those modeled in economics.

SIMULATING INTELLIGENCE – THE TURING TEST

Alan Turing, inventor of the Turing machine, an abstract concept that's used in algorithm theory, suggested a way to test intelligence. This test is referred to as the **Turing test** in AI literature.

Using a text interface, an interrogator chats to a human and a chatbot. The job of the chatbot is to mislead the interrogator to the extent that they cannot tell whether the computer is human.

WHAT DISCIPLINES DO WE NEED TO PASS THE TURING TEST?

First, we need to understand a spoken language to know what the interrogator is saying. We do this by using **Natural Language Processing** (**NLP**). We also must respond to the interrogator in a credible way by learning from previous questions and answers using AI models.

We need to be an expert of things that the human mind tends to be interested in. We need to build an expert system of humanity, involving the taxonomy of objects and abstract thoughts in our world, as well as historical events and even emotions.

Passing the Turing test is very hard. Current predictions suggest we won't be able to create a system good enough to pass the Turing test until the late 2020s. Pushing this even further, if this is not enough, we can advance to the Total Turing Test, which also includes movement and vision.

Next, we will move on and look at the tools and learning models in AI.

AI TOOLS AND LEARNING MODELS

In the previous sections, we discovered the fundamentals of AI. One of the core tasks of AI is learning. This is where intelligent agents come into the picture.

INTELLIGENT AGENTS

When solving AI problems, we create an actor in the environment that can gather data from its surroundings and influence its surroundings. This actor is called an **intelligent agent**.

An intelligent agent is as follows:

- Is autonomous

- Observes its surroundings through sensors

- Acts in its environment using actuators (which are the components that are responsible for moving and controlling a mechanism)

- Directs its activities toward achieving goals

Agents may also learn and have access to a knowledge base.

We can think of an agent as a function that maps perceptions to actions. If the agent has an internal knowledge base, then perceptions, actions, and reactions may alter the knowledge base as well.

Actions may be rewarded or punished. Setting up a correct goal and implementing a carrot and stick situation helps the agent learn. If goals are set up correctly, agents have a chance of beating the often more complex human brain. This is because the primary goal of the human brain is survival, regardless of the game we are playing. An agent's primary motive is reaching the goal itself. Therefore, intelligent agents do not get embarrassed when making a random move without any knowledge.

THE ROLE OF PYTHON IN AI

In order to put basic AI concepts into practice, we need a programming language that supports AI. In this book, we have chosen Python. There are a few reasons why Python is such a good choice for AI:

- **Convenience and Compatibility:** Python is a high-level programming language. This means that you don't have to worry about memory allocation, pointers, or machine code in general. You can write code in a convenient fashion and rely on Python's robustness. Python is also cross-platform compatible.

- **Popularity:** The strong emphasis on developer experience makes Python a very popular choice among software developers. In fact, according to a 2018 developer survey by https://www.hackerrank.com, across all ages, Python ranks as the number one preferred language of software developers. This is because Python is easily readable and simple. Therefore, Python is great for rapid application development.

- **Efficiency:** Despite being an interpreted language, Python is comparable to other languages that are used in data science, such as R. Its main advantage is memory efficiency, since Python can handle large, in-memory databases.

> **NOTE**
>
> Python is a multi-purpose language. It can be used to create desktop applications, database applications, mobile applications, and games.
> The network programming features of Python are also worth mentioning. Furthermore, Python is an excellent prototyping tool.

WHY IS PYTHON DOMINANT IN MACHINE LEARNING, DATA SCIENCE, AND AI?

To understand the dominant nature of Python in machine learning, data science, and AI, we have to compare Python to other languages that are also used in these fields.

Compared to R, which is a programming language built for statisticians, Python is much more versatile and easy as it allows programmers to build a diverse range of applications, from games to AI applications.

Compared to Java and C++, writing programs in Python is significantly faster. Python also provides a high degree of flexibility.

There are some languages that are similar in nature when it comes to flexibility and convenience: Ruby and JavaScript. Python has an advantage over these languages because of the AI ecosystem that's available for Python. In any field, open source, third-party library support vastly determines the success of that language. Python's third-party AI library support is excellent.

ANACONDA IN PYTHON

We installed Anaconda in the *Preface*. Anaconda will be our number one tool when it comes to experimenting with AI.

Anaconda comes with packages, IDEs, data visualization libraries, and high-performance tools for parallel computing in one place. Anaconda hides configuration problems and the complexity of maintaining a stack for data science, machine learning, and AI. This feature is especially useful in Windows, where version mismatches and configuration problems tend to arise the most.

Anaconda comes with Jupyter Notebook, where you can write code and comments in a documentation style. When you experiment with AI features, the flow of your ideas resembles an interactive tutorial where you run each step of your code.

> **NOTE**
>
> **IDE** stands for **Integrated Development Environment**. While a text editor provides some functionalities to highlight and format code, an IDE goes beyond the features of text editors by providing tools to automatically refactor, test, debug, package, run, and deploy code.

PYTHON LIBRARIES FOR AI

The list of libraries presented here is not complete as there are more than 700 available in Anaconda. However, these specific ones will get you off to a good start because they will give you a good foundation to be able to implement the fundamental AI algorithms in Python:

- **NumPy**: NumPy is a computing library for Python. As Python does not come with a built-in array data structure, we have to use a library to model vectors and matrices efficiently. In data science, we need these data structures to perform simple mathematical operations. We will use NumPy extensively in future chapters.

- **SciPy**: SciPy is an advanced library containing algorithms that are used for data science. It is a great complementary library to NumPy because it gives you all the advanced algorithms you need, whether it be a linear algebra algorithm, image processing tool, or a matrix operation.

- **pandas**: pandas provides fast, flexible, and expressive data structures, such as one-dimensional series and two-dimensional DataFrames. It efficiently loads, formats, and handles complex tables of different types.

- **scikit-learn**: scikit-learn is Python's main machine learning library. It is based on the NumPy and SciPy libraries. scikit-learn provides you with the functionality required to perform both classification and regression, data preprocessing, as well as supervised and unsupervised learning.

- **NLTK**: We will not deal with NLP in this book, but NLTK is still worth mentioning because this library is the main natural language toolkit of Python. You can perform classification, tokenization, stemming, tagging, parsing, semantic reasoning, and many other operations using this library.

- **TensorFlow**: TensorFlow is Google's neural network library, and it is perfect for implementing deep learning AI. The flexible core of TensorFlow can be used to solve a vast variety of numerical computation problems. Some real-world applications of TensorFlow include Google voice recognition and object identification.

A BRIEF INTRODUCTION TO THE NUMPY LIBRARY

The NumPy library will play a major role in this book, so it is worth exploring it further.

After launching your Jupyter Notebook, you can simply import **numpy** as follows:

```
import numpy as np
```

Once **numpy** has been imported, you can access it using its alias, **np**. NumPy contains the efficient implementation of some data structures, such as vectors and matrices.

Let's see how we can define vectors and matrices:

```
np.array([1,3,5,7])
```

The expected output is this:

```
array([1, 3, 5, 7])
```

We can declare a matrix using the following syntax:

```
A = np.mat([[1,2],[3,3]])
A
```

The expected output is this:

```
matrix([[1, 2],
        [3, 3]])
```

The **array** method creates an array data structure, while **.mat** creates a matrix.

We can perform many operations with matrices. These include addition, subtraction, and multiplication. Let's have a look at these operations here:

Addition in matrices:

```
A + A
```

The expected output is this:

```
matrix([[2, 4],
        [6, 6]])
```

Subtraction in matrices:

```
A - A
```

The expected output is this:

```
matrix([[0, 0],
        [0, 0]])
```

Multiplication in matrices:

```
A * A
```

The expected output is this:

```
matrix([[ 7,  8],
        [12, 15]])
```

Matrix addition and subtraction work cell by cell.

Matrix multiplication works according to linear algebra rules. To calculate matrix multiplication manually, you have to align the two matrices, as follows:

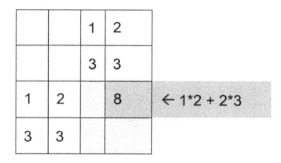

Figure 1.1: Multiplication calculation with two matrices

To get the *(i,j)*th element of the matrix, you compute the dot (scalar) product on the *i*th row of the matrix with the *j*th column. The scalar product of two vectors is the sum of the product of their corresponding coordinates.

Another frequent matrix operation is the determinant of the matrix. The determinant is a number associated with square matrices. Calculating the determinant using NumPy's **linalg** function (linear algebra algorithms) can be seen in the following line of code:

```
np.linalg.det( A )
```

The expected output is this:

```
-3.000000000000004
```

Technically, the determinant can be calculated as **1*3 − 2*3 = −3**. Notice that NumPy calculates the determinant using floating-point arithmetic, so the accuracy of the result is not perfect. The error is due to the way floating points are represented in most programming languages.

We can also transpose a matrix, as shown in the following line of code:

```
np.matrix.transpose(A)
```

The expected output is this:

```
matrix([[1, 3],
        [2, 3]])
```

When calculating the transpose of a matrix, we flip its values over its main diagonal.

NumPy has many other important features, so we will use it in most of the chapters in this book.

EXERCISE 1.01: MATRIX OPERATIONS USING NUMPY

We will be using Jupyter Notebook and the following matrix to solve this exercise.

We will calculate the square of the matrix, which is determinant of the matrix and the transpose of the matrix shown in the following figure, using NumPy:

$$A = \begin{bmatrix} 1 & 2 & 3 \\ 4 & 5 & 6 \\ 7 & 8 & 9 \end{bmatrix}$$

Figure 1.2: A simple matrix representation

The following steps will help you to complete this exercise:

1. Open a new Jupyter Notebook file.

2. Import the **numpy** library as **np**:

```
import numpy as np
```

3. Create a two-dimensional array called **A** for storing the **[[1,2,3],[4,5,6],[7,8,9]]** matrix using **np.mat**:

```
A = np.mat([[1,2,3],[4,5,6],[7,8,9]])
A
```

The expected output is this:

```
matrix([[1, 2, 3],
        [4, 5, 6],
        [7, 8, 9]])
```

> **NOTE**
>
> If you have created an **np.array** instead of **np.mat**, the solution for the array multiplication will be incorrect.

4. Next, we perform matrix multiplication using the asterisk and save the result in a variable called **matmult**, as shown in the following code snippet:

```
matmult = A * A
matmult
```

The expected output is this:

```
matrix([[ 30,  36,  42],
        [ 66,  81,  96],
        [102, 126, 150]])
```

5. Next, manually calculate the square of **A** by performing matrix multiplication. For instance, the top-left element of the matrix is calculated as follows:

```
1 * 1 + 2 * 4 + 3 * 7
```

The expected output is this:

```
30
```

6. Use **np.linalg.det** to calculate the determinant of the matrix and save the result in a variable called **det**:

```
det = np.linalg.det( A )
det
```

The expected output (might vary slightly) is this:

```
0.0
```

7. Use **np.matrix.transpose** to get the transpose of the matrix and save the result in a variable called **transpose**:

```
transpose = np.matrix.transpose(A)
transpose
```

The expected output is this:

```
matrix([[1, 4, 7],
        [2, 5, 8],
        [3, 6, 9]])
```

If **T** is the transpose of matrix **A**, then **T[j][i]** is equal to **A[i][j]**.

> **NOTE**
>
> To access the source code for this specific section, please refer to https://packt.live/316Vd6Z.
>
> You can also run this example online at https://packt.live/2BrogHL. You must execute the entire Notebook in order to get the desired result.

By completing this exercise, you have seen that NumPy comes with many useful features for vectors, matrices, and other mathematical structures.

In the upcoming section, we will be implementing AI in an interesting tic-tac-toe game using Python.

PYTHON FOR GAME AI

An AI game player is nothing but an intelligent agent with a clear goal: to win the game and defeat all the other players. AI experiments have achieved surprising results when it comes to games. Today, no human can defeat an AI in the game of chess.

The game *Go* was the last game where human players could consistently defeat a computer player. However, in 2017, Google's game-playing AI called AlphaGo defeated the world number 1 ranked Go player.

INTELLIGENT AGENTS IN GAMES

An intelligent agent plays according to the rules of the game. The agent can sense the current state of the game through its sensors and can evaluate the potential steps. Once the agent finds the best possible step, it performs the action using its actuators. The agent finds the best possible action to reach the goal based on the information it has. Actions are either rewarded or punished. The carrot and stick are excellent examples of rewards and punishment. Imagine a donkey in front of your cart. You put a carrot in front of the eyes of the donkey, so the animal starts walking toward it. As soon as the donkey stops, the rider may apply punishment with a stick. This is not a human way of moving, but rewards and punishment control living organisms to some extent. The same happens to humans at school, at work, and in everyday life as well. Instead of carrots and sticks, we have income and legal punishment to shape our behavior.

In most games, a good sequence of actions results in a reward. When a human player feels rewarded, that makes the human feel happy. Humans tend to act in a way that maximizes their happiness. Intelligent agents, on the other hand, are only interested in their goal, which is to maximize their reward and minimize the punishment that's affecting their performance score.

When modeling games, we must determine their state space. An action causes a state transition. When we explore the consequences of all possible actions, we get a decision tree. This tree goes deeper as we start exploring the possible future actions of all players until the game ends.

The strength of AI is the execution of millions of possible steps each second. Therefore, game AI often boils down to a search exercise. When exploring all of the possible sequences of moves in a game, we get the state tree of a game.

Consider a chess AI. What is the problem with evaluating all possible moves by building a state tree consisting of all of the possible sequences of moves?

Chess is an EXPTIME game complexity-wise. The number of possible moves explodes combinatorically.

White starts with 20 possible moves: the eight pawns may move either one or two steps, and the two knights may move either up-up-left, or up-up-right. Then, black can make any of these 20 moves. There are already 20*20 = 400 possible combinations after just one move per player.

After the second move, we get 8,902 possible board constellations, and this number just keeps on growing. Just take seven moves, and you have to search through 10,921,506 possible constellations.

The average length of a chess game is approximately 40 moves. Some exceptional games take more than 200 moves to finish.

As a consequence, the computer player simply does not have time to explore the whole state space. Therefore, the search activity has to be guided with proper rewards, punishment, and simplifications of the rules.

BREADTH FIRST SEARCH AND DEPTH FIRST SEARCH

Creating a game AI is often a search exercise. Therefore, we need to be familiar with the two primary search techniques:

- **Breadth First Search** (**BFS**)
- **Depth First Search** (**DFS**)

These search techniques are applied on a directed rooted tree.

A tree is a data structure that has **nodes**, and **edges** connecting these nodes in such a way that any two nodes of the tree are connected by exactly one path. Have a look at the following figure:

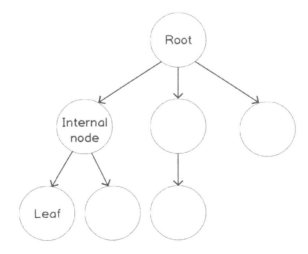

Figure 1.3: A directed rooted tree

When the tree is rooted, there is a special node in the tree called the **root**, which is where we begin our traversal. A directed tree is a tree where the edges may only be traversed in one direction. Nodes may be internal nodes or leaves. **Internal nodes** have at least one edge, through which we can leave the node. A **leaf** has no edges pointing out from the node.

In AI search, the root of the tree is the starting state. We traverse from this state by generating the successor nodes of the search tree. Search techniques differ, depending on the order in which we visit these successor nodes.

BREADTH FIRST SEARCH (BFS)

BFS is a search technique that, starting from the root node (node 1), will start exploring the closest node on the same depth (or level) before moving to the next depth:

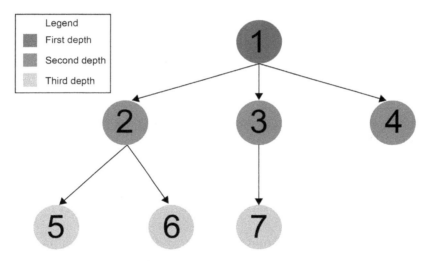

Figure 1.4: A BFS tree

In the preceding figure, you can see the search order of the BFS technique. Starting from the root node (**1**), BFS will go to the next level and explore the closest node (**2**) before looking at the other nodes on the same level (**3** and **4**). Then, it will move to the next level and explore **5** and **6** as they are close to each other before going back through to node **3**, finishing on the last node (**7**), and so on.

DEPTH FIRST SEARCH (DFS)

DFS is a search technique that, starting from the root node (node 1), will start exploring the same branch as much as possible before moving to the next closest branch:

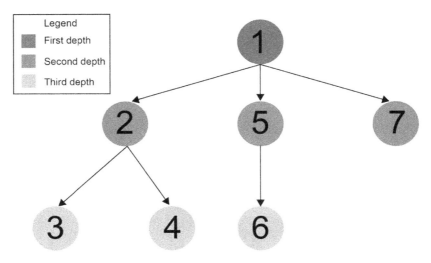

Figure 1.5: A DFS tree

In the preceding figure, you can see the search order of the DFS technique. Starting from the root node (**1**), DFS will go to the closest node (**2**) and explore all the way to the end of the branch (**3**) on the third depth before going back to the node (**2**) and finish by exploring its second branch (**4**). Then, it will move back to the second depth and start the same process with the next branch (**6**) before finishing with the last branch (**7**).

Now, suppose we have a tree defined by its root and a function that generates all the successor nodes from the root. In the following example, each node has a value and a depth. We start from **1** and may either increase the value by **1** or **2**. Our goal is to reach the value **5**:

NOTE

The code snippet shown here uses a backslash (\ ) to split the logic across multiple lines. When the code is executed, Python will ignore the backslash, and treat the code on the next line as a direct continuation of the current line.

```
root = {'value': 1, 'depth': 1}
def succ(node):
    if node['value'] == 5:
        return []
    elif node['value'] == 4:
        return [{'value': 5,'depth': node['depth']+1}]
    else:
        return [{'value': node['value']+1, \
                 'depth':node['depth']+1}, \
                {'value': node['value']+2, \
                 'depth':node['depth']+1}]
```

In the preceding code snippet, we have initialized the root node as having a value and depth of **1**. Then, we created a function called **succ** that takes a node as input. This function will have 3 different cases:

- If the input node value is **5**, then return nothing as we will have already reached our goal (**5**).

- If the input node value is **4**, then return the value **5** and add **1** to the current depth.

- If the value is anything else, then add **1** to the depth and create two cases for the value, **+1** and **+2**.

First, we will perform BFS, as shown here:

```
def bfs_tree(node):
    nodes_to_visit = [node]
    visited_nodes = []
    while len(nodes_to_visit) > 0:
        current_node = nodes_to_visit.pop(0)
        visited_nodes.append(current_node)
        nodes_to_visit.extend(succ(current_node))
    return visited_nodes

bfs_tree(root)
```

In the preceding code snippet, we have implemented the **bfs_tree** function by taking a node as input. This function can be broken down into three parts:

The **first part** is initializing the **nodes_to_visit** and **visited_nodes** variables.

The **second part** is where BFS is implemented:

- The **current_node** variable takes away the first element of the **nodes_to_visit** variable.

- The **visited_nodes** variable adds this element to its list.

- The **nodes_to_visit** variable adds the newly generated nodes from the call of **succ** with the **current_node** as input to it.

The preceding three instructions are wrapped into a loop defined by the number of elements inside the **nodes_to_visit** variable. As long as **nodes_to_visit** has at least one element, then the loop will keep going.

The **third part**, which is the end of the function, will return the entire list of values from the **visited_nodes** variable.

The expected output is this:

```
[{'value': 1, 'depth': 1},
 {'value': 2, 'depth': 2},
 {'value': 3, 'depth': 2},
 {'value': 3, 'depth': 3},
 {'value': 4, 'depth': 3},
 {'value': 4, 'depth': 3},
 {'value': 5, 'depth': 3},
 {'value': 4, 'depth': 4},
 {'value': 5, 'depth': 4},
 {'value': 5, 'depth': 4},
 {'value': 5, 'depth': 4},
 {'value': 5, 'depth': 5}]
```

As you can see, BFS is searching through the values of the same depth before moving to the next level of depth and exploring the values of it. Notice how depth and value are increasing in sequence. This will not be the case in DFS.

If we had to traverse a graph instead of a directed rooted tree, BFS would look different: whenever we visit a node, we would have to check whether the node had been visited before. If the node had been visited before, we would simply ignore it.

In this chapter, we will only use Breadth First Traversal on trees. DFS is surprisingly similar to BFS. The difference between DFS and BFS is the sequence in which you access the nodes. While BFS visits all the children of a node before visiting any other nodes, DFS digs deep into the tree.

Have a look at the following example, where we're implementing DFS:

```python
def dfs_tree(node):
    nodes_to_visit = [node]
    visited_nodes = []
    while len(nodes_to_visit) > 0:
        current_node = nodes_to_visit.pop()
        visited_nodes.append(current_node)
        nodes_to_visit.extend(succ(current_node))
    return visited_nodes


dfs_tree(root)
```

In the preceding code snippet, we have implemented the **dfs_tree** function by taking a node as input. This function can be broken down into three parts:

The **first part** is initializing the **nodes_to_visit** and **visited_nodes** variables.

The **second part** is where DFS is implemented:

- The **current_node** variable takes away the last element of the **nodes_to_visit** variable.

- The **visited_nodes** variable adds this element to its list.

- The **nodes_to_visit** variable adds the newly generated nodes from the call of **succ** with **current_node** as input to it.

The preceding three instructions are wrapped into a loop defined by the number of elements inside the **nodes_to_visit** variable. As long as **nodes_to_visit** has at least one element, then the loop will keep going.

At the end, that is at the **third part**, the function will return the entire list of values from **visited_nodes**.

As you can see, the main difference between BFS and DFS is the order in which we took an element out of **nodes_to_visit**. For BFS, we take the first element, whereas for DFS, we take the last one.

The expected output is this:

```
[{'value': 1, 'depth': 1},
 {'value': 3, 'depth': 2},
 {'value': 5, 'depth': 3},
 {'value': 4, 'depth': 3},
 {'value': 5, 'depth': 4},
 {'value': 2, 'depth': 2},
 {'value': 4, 'depth': 3},
 {'value': 5, 'depth': 4},
 {'value': 3, 'depth': 3},
 {'value': 5, 'depth': 4},
 {'value': 4, 'depth': 4},
 {'value': 5, 'depth': 5}]
```

Notice how the DFS algorithm digs deep fast (the depth reaches higher values faster than BFS). It does not necessarily find the shortest path first, but it is guaranteed to find a leaf before exploring a second path.

In game AI, the BFS algorithm is often better for the evaluation of game states because DFS may get lost. Imagine starting a chess game, where a DFS algorithm may easily get lost in exploring the options for a move.

EXPLORING THE STATE SPACE OF A GAME

Let's explore the state space of a simple game: tic-tac-toe. A state space is the set of all possible configurations of a game, which, in this case, means all the possible moves.

In tic-tac-toe, a 3x3 game board is given. Two players play this game. One plays with the sign X, while the other plays with the sign O. X starts the game, and each player makes a move after the other. The goal of the game is to get three of your own signs horizontally, vertically, or diagonally.

Let's denote the cells of the tic-tac-toe board, as follows:

1	2	3
4	5	6
7	8	9

Figure 1.6: Tic-tac-toe board

In the following example, **X** started at position **1**. **O** retaliated at position **5**, **X** made a move at position **9**, and then **O** moved to position **3**:

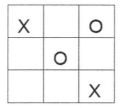

Figure 1.7: Tic-tac-toe board with noughts and crosses

This was a mistake by the second player, because now **X** is forced to place a sign on cell **7**, creating two future scenarios for winning the game. It does not matter whether **O** defends by moving to cell **4** or **8** – **X** will win the game by selecting the other unoccupied cell.

> **NOTE**
>
> You can try out the game at http://www.half-real.net/tictactoe/.

For simplicity, we will only explore the state space belonging to the cases when the AI player starts. We will start with an AI player who plays randomly, placing a sign in an empty cell. After playing with this AI player, we will create a complete decision tree. Once we generate all possible game states, you will experience their combinatorial explosion. As our goal is to make these complexities simple, we will use several different techniques to make the AI player smarter and reduce the size of the decision tree.

By the end of this experiment, we will have a decision tree that has fewer than 200 different game endings and, as a bonus, the AI player will never lose a single game.

To make a random move, you will have to know how to choose a random element from a list using Python. We will use the **choice** function of the **random** library to do so:

```
from random import choice
choice([2, 4, 6, 8])
```

This time, the output is **6**, but for you, it can be any number from the list.

The output of the **choice** function is a random element of the list.

> **NOTE**
>
> We will use the factorial notation in the following section. A factorial is denoted by the "!" exclamation mark. By definition, 0! = 1, and n! = n*(n-1)!. In our example, 9! = 9* 8! = 9*8*7! = … = 9*8*7*6*5*4*3*2*1.

ESTIMATING THE NUMBER OF POSSIBLE STATES IN A TIC-TAC-TOE GAME

Make a rough estimate of the number of possible states on each level of the state space of the tic-tac-toe game:

- In our estimation, we will not stop until all of the cells of the board have been filled. A player might win before the game ends, but for the sake of uniformity, we will continue the game.

- The first player will choose one of the nine cells. The second player will choose one out of the eight remaining cells. The first player can then choose one out of the seven remaining cells. This goes on until either player wins the game, or the first player is forced to make the ninth and last move.

- The number of possible decision sequences is, therefore, 9! = 362,880. A few of these sequences are invalid because a player may win the game in fewer than nine moves. It takes at least five moves to win a game because the first player needs to move three times.

- To calculate the exact size of the state space, we have to calculate the number of games that are won in five, six, seven, and eight steps. This calculation is simple, but due to its brute-force nature, it is beyond our scope. Therefore, we will settle on the magnitude of the state space.

> **NOTE**
>
> After generating all possible tic-tac-toe games, researchers counted 255,168 possible games. Out of those games, 131,184 were won by the first player, 77,904 were won by the second player, and 46,080 games ended with a draw. Visit http://www.half-real.net/tictactoe/allgamesoftictactoe. zip to download all possible tic-tac-toe games.

Even a simple game such as tic-tac-toe has a lot of states. Just imagine how hard it would be to start exploring all possible chess games. Therefore, we can conclude that brute-force searching is rarely ideal.

EXERCISE 1.02: CREATING AN AI WITH RANDOM BEHAVIOR FOR THE TIC-TAC-TOE GAME

In this exercise, we'll create a framework for the tic-tac-toe game for experimentation. We will be modeling the game on the assumption that the AI player always starts the game. You will create a function that prints your internal representation, allows your opponent to enter a move randomly, and determines whether a player has won.

> **NOTE**
>
> To ensure that this happens correctly, you will need to have completed the previous exercise.

The following steps will help you to complete this exercise:

1. Begin by opening a new Jupyter Notebook file.

2. We will import the **choice** function from the **random** library:

```
from random import choice
```

3. Now, model the nine cells in a simple string for simplicity. A nine-character long Python string stores these cells in the following order: "**123456789**". Let's determine the index triples that must contain matching signs so that a player wins the game:

```
combo_indices = [[0, 1, 2], [3, 4, 5], [6, 7, 8], [0, 3, 6], \
                 [1, 4, 7], [2, 5, 8], [0, 4, 8], [2, 4, 6]]
```

4. Define the sign constants for empty cells, the AI, and the opponent player:

```
EMPTY_SIGN = '.'
AI_SIGN = 'X'
OPPONENT_SIGN = 'O'
```

In the preceding code snippet, we have assigned a different sign for the AI and the player.

5. Create a function that prints a board. We will add an empty row before and after the board so that we can easily read the game state:

```
def print_board(board):
    print(" ")
    print(' '.join(board[:3]))
    print(' '.join(board[3:6]))
    print(' '.join(board[6:]))
    print(" ")
```

6. Describe a move of the human player. The input arguments are the boards, the row numbers from **1** to **3**, and the column numbers from **1** to **3**. The return value of this function is a board containing the new move:

```
def opponent_move(board, row, column):
    index = 3 * (row - 1) + (column - 1)
    if board[index] == EMPTY_SIGN:
        return board[:index] + OPPONENT_SIGN + board[index+1:]
    return board
```

Here, we have defined a function called **opponent_move** that will help us to calculate the index of the board based on the input (row and column). You will be able to see the resulting position on the board.

7. Now, we need to define a random move on the part of the AI player. We will generate all possible moves with the **all_moves_from_board** function, and then select a random move from the list of possible moves:

```
def all_moves_from_board(board, sign):
    move_list = []
    for i, v in enumerate(board):
        if v == EMPTY_SIGN:
            move_list.append(board[:i] + sign + board[i+1:])
    return move_list

def ai_move(board):
    return choice(all_moves_from_board(board, AI_SIGN))
```

In the preceding code snippet, we defined a function called **all_moves_from_board** that goes through all the indexes on the board and checks whether they are empty (**v == EMPTY_SIGN**). If that's the case, this means that the move can be played and that the index has been added to a list of moves (**move_list**). Finally, we defined the **ai_move** function in order to randomly let the AI choose an index that is equal to a move in the game.

8. Determine whether a player has won the game:

```
def game_won_by(board):
    for index in combo_indices:
        if board[index[0]] == board[index[1]] == \
            board[index[2]] != EMPTY_SIGN:
            return board[index[0]]
    return EMPTY_SIGN
```

In the preceding code snippet, we have defined the **game_won_by** function, which checks whether the board contains a combo of three identical indexes from the **combo_indices** variable to end the game.

9. Finally, create a game loop so that we can test the interaction between the computer player and the human player. We will conduct a brute-force search in the following examples:

```
def game_loop():
    board = EMPTY_SIGN * 9
    empty_cell_count = 9
    is_game_ended = False
    while empty_cell_count > 0 and not is_game_ended:
        if empty_cell_count % 2 == 1:
            board = ai_move(board)
        else:
            row = int(input('Enter row: '))
            col = int(input('Enter column: '))
            board = opponent_move(board, row, col)
        print_board(board)
        is_game_ended = game_won_by(board) != EMPTY_SIGN
        empty_cell_count = sum(1 for cell in board \
                               if cell == EMPTY_SIGN)
    print('Game has been ended.')
```

In the preceding code snippet, we defined the function, which can be broken down into various parts.

The first part is to initialize the board and fill it with empty signs (**board = EMPTY_SIGN * 9**). Then, we create a counter of the empty cell, which will help us to create a loop and determine the AI's turn.

The second part is to create a function for the player and the AI engine to play the game against each other. As soon as one player makes a move, the **empty_cell_count** variable will decrease by 1. The loop will keep going until either the **game_won_by** function finds a winner or there are no more possible moves on the board.

10. Use the **game_loop** function to run the game:

```
game_loop()
```

The expected output (partially shown) is this:

```
.  O  X
X  O  .
.  X  .

Enter  row:  3
Enter  column:  3

.  O  X
X  O  .
.  X  O


.  O  X
X  O  X
.  X  O

Enter  row:  1
Enter  column:  1

O  O  X
X  O  X
.  X  O

Game  has  been  ended.
```

Figure 1.8: Final output (partially shown) of the game

NOTE

To access the source code for this specific section, please refer to
https://packt.live/3fUws2l.

You can also run this example online at https://packt.live/3hVzjcT. You must
execute the entire Notebook in order to get the desired result.

By completing this exercise, you have seen that even an opponent who is playing
randomly may win from time to time if their opponent makes a mistake.

ACTIVITY 1.01: GENERATING ALL POSSIBLE SEQUENCES OF STEPS IN A TIC-TAC-TOE GAME

This activity will explore the combinatorial explosion that is possible when two players play randomly. We will be using a program that, building on the previous results, generates all possible sequences of moves between a computer player and a human player.

Let's assume that the human player may make any possible move. In this example, given that the computer player is playing randomly, we will examine the wins, losses, and draws belonging to two randomly playing players:

The following steps will help you to complete this activity:

1. Reuse all the function codes of *Steps 2–9* from the previous *Exercise 1.02, Creating an AI with Random Behavior for the Tic-Tac-Toe Game*.

2. Create a function that maps the **all_moves_from_board** function on each element of a list of board spaces/squares. This way, we will have all of the nodes of a decision tree. The decision tree starts with **[EMPTY_SIGN * 9]** and expands after each move.

3. Create a **filter_wins** function that takes finished games out of the list of moves and appends them in an array containing the board states won by the AI player and the opponent player.

4. Create a **count_possibilities** function that prints and returns the number of decision tree leaves that ended with a draw, were won by the first player, and were won by the second player:

5. We have up to nine steps in each state. In the 0th, 2nd, 4th, 6th, and 8th iterations, the AI player moves. In all other iterations, the opponent moves. We create all possible moves in all steps and take out finished games from the move list.

6. Finally, execute the number of possibilities to experience the combinatorial explosion.

The expected output is this:

```
step 0. Moves: 1
step 1. Moves: 9
step 2. Moves: 72
step 3. Moves: 504
step 4. Moves: 3024
step 5. Moves: 13680
step 6. Moves: 49402
step 7. Moves: 111109
step 8. Moves: 156775
First player wins: 106279
Second player wins: 68644
Draw 91150
Total 266073
```

> **NOTE**
>
> The solution to this activity can be found on page 322.

So far, we've understood the significance of an intelligent agent. We also examined the game states for a game AI. Now, we will focus on how to create and introduce intelligence to an agent.

We will look at reducing the number of states in the state space, analyze the stages that a game board can undergo, and make the environment work in such a way that we win.

Have a look at the following exercise, where we'll teach an intelligent agent to win.

EXERCISE 1.03: TEACHING THE AGENT TO WIN

In this exercise, we will see how the steps needed to win can be reduced. We will be making the agent that we developed in the previous section activity detect situations where it can win a game.

The following steps will help you to complete this exercise:

1. Open a new Jupyter Notebook file.

2. Reuse the previous code from *Steps 2–6* from *Activity 1, Generating All Possible Sequences of Steps in a Tic-Tac-Toe Game*.

3. Define two functions, **ai_move** and **all_moves_from_board**.

   We create **ai_move** so that it returns a move that will consider its own previous moves. If the game can be won in that move, **ai_move** will select that move:

   ```
   def ai_move(board):
       new_boards = all_moves_from_board(board, AI_SIGN)
       for new_board in new_boards:
           if game_won_by(new_board) == AI_SIGN:
               return new_board
       return choice(new_boards)
   ```

   In the preceding code snippet, we have defined the **ai_move** function, which will make the AI choose a winning move from a list of all the possible moves from the current state of the game if it's applicable. If not, it will still choose a random move.

4. Next, test the code snippet with a game loop. Whenever the AI has the opportunity to win the game, it will always place the **X** in the correct cell:

   ```
   game_loop()
   ```

The expected output is this:

```
X . .

. . .

. . .


Enter row: 1
Enter column: 2

X O .

. . .

. . .


X O .

. . .

. . X


Enter row: 1
Enter column: 3

X O O

. . .

. . X


X O O

. X .

. . X


Game has been ended.
```

Figure 1.9: The agent winning the game

5. Now, count all the possible moves. To do this, we must change the **all_moves_from_board** function to include this improvement. We must do this so that, if the game is won by **AI_SIGN**, it will return that value:

```
def all_moves_from_board(board, sign):
    move_list = []
    for i, v in enumerate(board):
        if v == EMPTY_SIGN:
            new_board = board[:i] + sign + board[i+1:]
            move_list.append(new_board)
            if game_won_by(new_board) == AI_SIGN:
                return [new_board]
    return move_list
```

In the preceding code snippet, we have defined a function to generate all possible moves. As soon as we find a move that wins the game for the AI, we return it. We do not care whether the AI has multiple options to win the game in one move – we just return the first possibility. If the AI cannot win, we return all possible moves. Let's see what this means in terms of counting all of the possibilities at each step.

6. Enter the following function to find all the possibilities.

```
first_player, second_player, \
draw, total = count_possibilities()
```

The expected output is this:

```
step 0. Moves: 1
 step 1. Moves: 9
 step 2. Moves: 72
 step 3. Moves: 504
 step 4. Moves: 3024
 step 5. Moves: 8525
 step 6. Moves: 28612
 step 7. Moves: 42187
 step 8. Moves: 55888
 First player wins: 32395
 Second player wins: 23445
 Draw 35544
 Total 91384
```

With that, we have seen that the AI is still not winning most of the time. This means that we need to introduce more concepts to the AI to make it stronger. To teach the AI how to win, we need to teach it how to make defensive moves against losses.

DEFENDING THE AI AGAINST LOSSES

In the next activity, we will make the AI computer player play better compared to our previous exercise so that we can reduce the state space and the number of losses.

ACTIVITY 1.02: TEACHING THE AGENT TO REALIZE SITUATIONS WHEN IT DEFENDS AGAINST LOSSES

In this activity, we will force the computer to defend against a loss if the player puts their third sign in a row, column, or diagonal line:

1. Reuse all the code from *Steps 2–6* from the previous, *Exercise 1.03*, *Teaching the Agent to Win*.

2. Create a function called **player_can_win** that takes all the moves from the board using the **all_moves_from_board** function and iterates over it using a variable called **next_move**. On each iteration, it checks whether the game can be won by the sign, and then it returns **true** or **false**.

3. Extend the AI's move so that it prefers making safe moves. A move is safe if the opponent cannot win the game in the next step.

4. Test the new application. You will find that the AI has made the correct move.

5. Place this logic in the state space generator and check how well the computer player is doing by generating all possible games.

The expected output is this:

```
step 0. Moves: 1
step 1. Moves: 9
step 2. Moves: 72
step 3. Moves: 504
step 4. Moves: 3024
step 5. Moves: 5197
step 6. Moves: 18606
step 7. Moves: 19592
step 8. Moves: 30936
First player wins: 20843
Second player wins: 962
Draw 20243
Total 42048
```

> **NOTE**
>
> The solution to this activity can be found on page 325.

Once we complete this activity, we notice that despite our efforts to make the AI better, it can still lose in **962** ways. We will eliminate all these losses in the next activity.

ACTIVITY 1.03: FIXING THE FIRST AND SECOND MOVES OF THE AI TO MAKE IT INVINCIBLE

In this activity, we will be combining our previous activities by teaching the AI how to recognize both a win and a loss so that it can focus on finding moves that are more useful than others. We will be reducing the possible games by hardcoding the first and second moves:

1. Reuse the code from *Steps 2–4* of the previous, *Activity 1.02, Teaching the Agent to Realize Situations When It Defends Against Losses*.

2. Count the number of empty fields on the board and make a hardcoded move in case there are 9 or 7 empty fields. You can experiment with different hardcoded moves.

3. Occupying any corner, and then occupying the opposite corner, leads to no losses. If the opponent occupies the opposite corner, making a move in the middle results in no losses.

4. After fixing the first two steps, we only need to deal with 8 possibilities instead of 504. We also need to guide the AI into a state where the hardcoded rules are enough so that it never loses a game.

The expected output is this:

```
step 0. Moves: 1
step 1. Moves: 1
step 2. Moves: 8
step 3. Moves: 8
step 4. Moves: 48
step 5. Moves: 38
step 6. Moves: 108
step 7. Moves: 76
step 8. Moves: 90
First player wins: 128
Second player wins: 0
Draw 60
Total 188
```

NOTE

The solution to this activity can be found on page 328.

Let's summarize the important techniques that we applied to reduce the state space so far:

- **Empirical simplification**: We accepted that the optimal first move is a corner move. We simply hardcoded a move instead of considering alternatives to focus on other aspects of the game. In more complex games, empirical moves are often misleading. The most famous chess AI victories often contain a violation of the common knowledge of chess grand masters.

- **Symmetry**: After we started with a corner move, we noticed that positions 1, 3, 7, and 9 were equivalent to the perspective of winning the game. Even though we didn't take this idea further, we noticed that we could even rotate the table to reduce the state space even further and consider all four corner moves as the exact same move.

- **Reduction in different permutations leading to the same state**: Suppose we can make the moves A or B and suppose our opponent makes move X, where X is not equal to either move A or B. If we explore the sequence A, X, B, and we start exploring the sequence B, X, then we don't have to consider the sequence B, X, A. This is because the two sequences lead to the exact same game state, and we have already explored a state containing these three moves before, that is, A, X, and B. The order of the sequence doesn't matter as it leads to the same result. This allows us to significantly reduce the number of possible moves.

- **Forced moves for the player**: When a player collects two signs horizontally, vertically, or diagonally, and the third cell in the row is empty, we are forced to occupy that empty cell either to win the game or to prevent the opponent from winning the game. Forced moves may imply other forced moves, which reduces the state space even further.

- **Forced moves for the opponent**: When a move from the opponent is clearly optimal, it does not make sense to consider scenarios where the opponent does not make the optimal move. When the opponent can win the game by occupying a cell, it does not matter whether we go on a long exploration of the cases when the opponent misses the optimal move. We save a lot less by not exploring cases when the opponent fails to prevent us from winning the game. This is because after the opponent makes a mistake, we will simply win the game.

- **Random move**: When we cannot decide and don't have the capacity to search, we move randomly. Random moves are almost always inferior to a search-based educated guess, but at times, we have no other choice.

HEURISTICS

In this section, we will formalize informed search techniques by defining and applying heuristics to guide our search. We will be looking at heuristics and creating them in the sections ahead.

UNINFORMED AND INFORMED SEARCHES

In the tic-tac-toe example, we implemented a greedy algorithm that first focused on winning, and then focused on not losing. When it comes to winning the game immediately, the greedy algorithm is optimal because there is never a better step than winning the game. When it comes to not losing, it matters how we avoid the loss. Our algorithm simply choses a random safe move without considering how many winning opportunities we have created.

BFS and DFS are part of uninformed searching because they consider all possible states in the game, which can be very time-consuming. On the other hand, heuristic informed searches will explore the space of available states intelligently in order to reach the goal faster.

CREATING HEURISTICS

If we want to make better decisions, we apply heuristics to guide the search in the right direction by considering long-term benefits. This way, we can make a more informed decision in the present based on what could happen in the future. This can also help us solve problems faster.

We can construct heuristics as follows:

- In terms of the utility of making a move in the game
- In terms of the utility of a given game state from the perspective of a player
- In terms of the distance from our goal

Heuristics are functions that evaluate a game state or a transition to a new game state based on their utility. Heuristics are the cornerstones of making a search problem informed.

In this book, we will use utility and cost as negated terms. Maximizing utility and minimizing the cost of a move are considered synonyms.

A commonly used example of a heuristic evaluation function occurs in pathfinding problems. Suppose we are looking to reach a destination or a goal. Each step has an associated cost symbolizing the travel distance. Our goal is to minimize the cost of reaching the destination or goal (minimizing the travel distance).

One example of heuristic evaluation for solving this pathfinding problem will be to take the coordinates between the current state (position) and the goal (destination) and calculate the distance between these two points. The distance between two points is the length of the straight line connecting the points. This heuristic is called the **Euclidean distance** (as shown in the *Figure 1.10*).

Now, suppose we define our pathfinding problem in a maze, where we can only move up, down, left, or right. There are a few obstacles in the maze that block our moves, so using the Euclidean distance is not ideal. A better heuristic would be to use the Manhattan distance, which can be defined as the sum of the horizontal and vertical distances between the coordinates of the current state and the goal.

ADMISSIBLE AND NON-ADMISSIBLE HEURISTICS

The two heuristics we just defined regarding pathfinding problems are called admissible heuristics when they're used on their given problem domain.

Admissible means that we may underestimate the cost of reaching the end state but that we never overestimate it. Later, we will explore an algorithm that finds the shortest path between the current state and the goal state. The optimal nature of this algorithm depends on whether we can define an admissible heuristic function.

An example of a non-admissible heuristic would be the Euclidean distance that's applied to a real-world map.

Imagine that we want to move from point A to point B in the city of Manhattan. Here, the Euclidean distance will be the straight line between the two points, but, as we know, we cannot just go straight in a city such as Manhattan (*unless we can fly*). In this case, the Euclidean distance is underestimating the cost of reaching the goal. A better heuristic would be the Manhattan distance:

Figure 1.10: Euclidian distance (blue line) and Manhattan distance (red line)
in the city of Manhattan

> **NOTE**
>
> The preceding map of Manhattan is sourced from Google Maps.

Since we overestimated the cost of traveling from the current node to the goal, the Euclidean distance is not admissible when we cannot move diagonally.

HEURISTIC EVALUATION

We can create a heuristic evaluation for our tic-tac-toe game state from the perspective of the starting player by defining the utility of a move.

HEURISTIC 1: SIMPLE EVALUATION OF THE ENDGAME

Let's define a simple heuristic by evaluating a board. We can set the utility for the game as one of the following:

- +1, if the state implies that the AI player will win the game

- -1, if the state implies that the AI player will lose the game

- 0, if a draw has been reached or no clear winner can be identified from the current state

This heuristic is simple because anyone can look at a board and analyze whether a player is about to win.

The utility of this heuristic depends on whether we can play many moves in advance. Notice that we cannot even win the game within five steps. In *Activity 1.01, Generating All Possible Sequences of Steps in a Tic-Tac-Toe Game*, we saw that by the time we reach step five, we have 13,680 possible combinations leading to it. In most of these 13,680 cases, our heuristic returns zero as we can't identify a clear winner yet.

If our algorithm does not look deeper than these five steps, we are completely clueless on how to start the game. Therefore, we should invent a better heuristic.

HEURISTIC 2: UTILITY OF A MOVE

Let's change the utility for the game as follows:

- Two AI signs in a row, column, or diagonal, and the third cell is empty: +1000 for the empty cell.

- The opponent has two signs in a row, column, or diagonal, and the third cell is empty: +100 for the empty cell.

- One AI sign in a row, column, or diagonal, and the other two cells are empty: +10 for the empty cells.

- No AI or opponent signs in a row, column, or diagonal: +1 for the empty cells.

- Occupied cells get a value of minus infinity. In practice, due to the nature of the rules, -1 will also do.

Why do we use a multiplicative factor of 10 for the first three rules compared to the fourth one? We do this because there are eight possible ways of making three in a row, column, and diagonal. So, even by knowing nothing about the game, we are certain that a lower-level rule may not accumulate to overriding a higher-level rule. In other words, we will never defend against the opponent's moves if we can win the game.

> **NOTE**
>
> As the job of our opponent is also to win, we can compute this heuristic from the opponent's point of view. Our task is to maximize this value, too, so that we can defend against the optimal plays of our opponent. This is the idea behind the Minmax algorithm as well, which will be covered later in this chapter. If we wanted to convert this heuristic into a heuristic that describes the current board, we could compute the heuristic value for all open cells and take the maximum of the values for the AI character so that we can maximize our utility.

For each board, we will create a utility matrix.

For example, consider the following board, with **O** signs as the player and **X** signs as the AI:

Figure 1.11: Tic-tac-toe game state

From here, we can construct its utility matrix shown in the following figure:

-1	-1	110
0	-1	10
-1	-1	-1

Figure 1.12: Tic-tac-toe game utility matrix

On the second row, the left cell is not beneficial if we were to select it. Note that if we had a more optimal utility function, we would reward blocking the opponent.

The two cells of the third column both get a **10**-point boost for two in a row.

The top-right cell also gets **100** points for defending against the diagonal of the opponent.

From this matrix, evidently, we should choose the top-right move. At any stage of the game, we were able to define the utility of each cell; this was a static evaluation of the heuristic function.

We can use this heuristic to guide us toward an optimal next move or to give a more educated score on the current board by taking the maximum of these values. We have technically used parts of this heuristic in the form of hardcoded rules. Note, though, that the real utility of heuristics is not the static evaluation of a board, but the guidance it provides for limiting the search space.

EXERCISE 1.04: TIC-TAC-TOE STATIC EVALUATION WITH A HEURISTIC FUNCTION

In this exercise, you will be performing a static evaluation on the tic-tac-toe game using a heuristic function.

The following steps will help you to complete this exercise:

1. Open a new Jupyter Notebook file.

2. Reuse the code from *Steps 2–6* of *Activity 1.01, Generating All Possible Sequences of Steps in a Tic-Tac-Toe Game*.

3. Create a function that takes the board as input and returns **0** if the cell is empty, and **−1** if it's not empty:

```
def init_utility_matrix(board):
    return [0 if cell == EMPTY_SIGN \
            else -1 for cell in board]
```

4. Next, create a function that takes the utility vector of possible moves, takes three indices inside the utility vector representing a triple, and returns a function, as shown in the following code snippet:

```
def generate_add_score(utilities, i, j, k):
    def add_score(points):
        if utilities[i] >= 0:
            utilities[i] += points
        if utilities[j] >= 0:
            utilities[j] += points
        if utilities[k] >= 0:
            utilities[k] += points
    return add_score
```

In the preceding code snippet, the returned function will expect a **points** parameter and the **utilities** vector as input and will add points to each cell in (**i**, **j**, **k**), as long as the original value of that cell is non-negative (**>=0**). In other words, we increased the utility of empty cells only.

5. Now, create the utility matrix belonging to any board constellation where you will add the **generate_add_score** function defined previously to update the score. You will also implement the rules that we discussed prior to this activity. These rules are as follows:

Two AI signs in a row, column, or diagonal, and the third cell is empty: +1000 for the empty cell.

The opponent has two signs in a row, column, or diagonal, and the third cell is empty: +100 for the empty cell.

One AI sign in a row, column, or diagonal, and the other two cells are empty: +10 for the empty cells.

No AI or opponent signs in a row, column, or diagonal: +1 for the empty cells.

Let's create the utility matrix now:

```
def utility_matrix(board):
    utilities = init_utility_matrix(board)
    for [i, j, k] in combo_indices:
        add_score = generate_add_score(utilities, i, j, k)
        triple = [board[i], board[j], board[k]]
        if triple.count(EMPTY_SIGN) == 1:
            if triple.count(AI_SIGN) == 2:
                add_score(1000)
            elif triple.count(OPPONENT_SIGN) == 2:
                add_score(100)
        elif triple.count(EMPTY_SIGN) == 2 and \
                        triple.count(AI_SIGN) == 1:
            add_score(10)
        elif triple.count(EMPTY_SIGN) == 3:
            add_score(1)
    return utilities
```

6. Create a function that selects the move with the highest utility value. If multiple moves have the same utility, the function returns both moves:

```
def best_moves_from_board(board, sign):
    move_list = []
    utilities = utility_matrix(board)
    max_utility = max(utilities)
    for i, v in enumerate(board):
        if utilities[i] == max_utility:
            move_list.append(board[:i] \
                            + sign \
                            + board[i+1:])
    return move_list


def all_moves_from_board_list(board_list, sign):
    move_list = []
```

```
get_moves = best_moves_from_board if sign \
            == AI_SIGN else all_moves_from_board
for board in board_list:
    move_list.extend(get_moves(board, sign))
return move_list
```

7. Now, run the application, as shown in the following code snippet:

```
first_player, second_player, \
draw, total = count_possibilities()
```

The expected output is this:

```
step 0. Moves: 1
step 1. Moves: 1
step 2. Moves: 8
step 3. Moves: 24
step 4. Moves: 144
step 5. Moves: 83
step 6. Moves: 214
step 7. Moves: 148
step 8. Moves: 172
First player wins: 504
Second player wins: 12
Draw 91
Total 607
```

> **NOTE**
>
> To access the source code for this specific section, please refer to https://packt.live/2VpGyAv.
>
> You can also run this example online at https://packt.live/2YnyO3K. You must execute the entire Notebook in order to get the desired result.

By completing this exercise, we have observed that the AI is underperforming compared to our previous activity, *Activity 1.03, Fixing the First and Second Moves of the AI to Make It Invincible*. In this situation, hardcoding the first two moves was better than setting up the heuristic, but this is because we haven't set up the heuristic properly.

USING HEURISTICS FOR AN INFORMED SEARCH

We have not experienced the real power of heuristics yet as we made moves without the knowledge of the effects of our future moves, thereby effecting reasonable play from our opponents.

Therefore, a more accurate heuristic leads to more losses than simply hardcoding the first two moves in the game. Note that in the previous section, we selected these two moves based on the statistics we generated based on running the game with fixed first moves. This approach is essentially what heuristic search should be all about.

TYPES OF HEURISTICS

Static evaluation cannot compete with generating hundreds of thousands of future states and selecting a play that maximizes our rewards. This is because our heuristics are not exact and are likely not admissible either.

We saw in the preceding exercise that heuristics are not always optimal. We came up with rules that allowed the AI to always win the game or finish with a draw. These heuristics allowed the AI to win very frequently, at the expense of losing in a few cases. A heuristic is said to be admissible if we underestimate the utility of a game state, but we never overestimate it.

In the tic-tac-toe example, we likely overestimated the utility in a few game states, and why is that? Because we ended up with a loss 12 times. A few of the game states that led to a loss had a maximum heuristic score. To prove that our heuristic is not admissible, all we need to do is find a potentially winning game state that we ignored while choosing a game state that led to a loss.

There are two more features that describe heuristics, that is, optimal and complete:

- Optimal heuristics always find the best possible solution.

- Complete heuristics has two definitions, depending on how we define the problem domain. In a loose sense, a heuristic is said to be complete if it always finds a solution. In a strict sense, a heuristic is said to be complete if it finds all possible solutions. Our tic-tac-toe heuristic is not complete because we ignored many possible winning states on purpose, favoring a losing state.

As you can see, defining an accurate heuristic requires a lot of details and thinking in order to obtain a perfect AI agent. If you are not correctly estimating the utility in the game states, then you can end up with an AI underperforming hardcoded rules.

In the next section, we'll look at a better approach to executing the shortest pathfinding between the current state and the goal state.

PATHFINDING WITH THE A* ALGORITHM

In the first two sections, we learned how to define an intelligent agent and how to create a heuristic that guides the agent toward a desired state. We learned that this was not perfect because, at times, we ignored a few winning states in favor of a few losing states.

Now, we will learn about a structured and optimal approach so that we can execute a search for finding the shortest path between the current state and the goal state by using the A* ("*A star*" instead of "*A asterisk*") algorithm.

Have a look at the following figure:

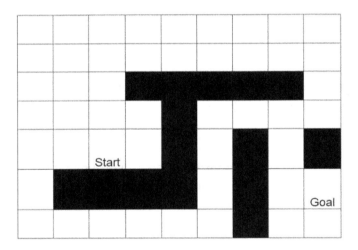

Figure 1.13: Finding the shortest path in a maze

For a human, it is simple to find the shortest path by merely looking at the figure. We can conclude that there are two potential candidates for the shortest path: route one starts upward, and route two starts to the left. However, the AI does not know about these options. In fact, the most logical first step for a computer player would be moving to the square denoted by the number **3** in the following figure.

Why? Because this is the only step that decreases the distance between the starting state and the goal state. All the other steps initially move away from the goal state:

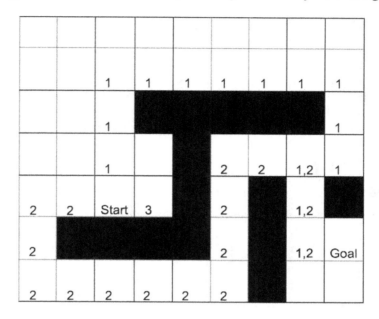

Figure 1.14: Shortest pathfinding game board with utilities

In the next exercise, we'll see how the BFS algorithm performs on the pathfinding problem before introducing you to the A* algorithm.

EXERCISE 1.05: FINDING THE SHORTEST PATH USING BFS

In this exercise, we will be finding the shortest path to our goal using the BFS algorithm.

The following steps will help you to complete this exercise:

1. Open a new Jupyter Notebook file.

2. Begin by importing the **math** library:

```
import math
```

3. Next, describe the board, the initial state, and the final state using Python. Create a function that returns a list of possible successors. Use tuples, where the first coordinate denotes the row number from **1** to **7** and the second coordinate denotes the column number from **1** to **9**:

```
size = (7, 9)
start = (5, 3)
end = (6, 9)
obstacles = {(3, 4), (3, 5), (3, 6), (3, 7), (3, 8), \
             (4, 5), (5, 5), (5, 7), (5, 9), (6, 2), \
             (6, 3), (6, 4), (6, 5), (6, 7),(7, 7)}
```

4. Next, use array comprehension to generate the successor states, as shown in the following code:

```
def successors(state, visited_nodes):
    (row, col) = state
    (max_row, max_col) = size
    succ_states = []
    if row > 1:
        succ_states += [(row-1, col)]
    if col > 1:
        succ_states += [(row, col-1)]
    if row < max_row:
        succ_states += [(row+1, col)]
    if col < max_col:
        succ_states += [(row, col+1)]
    return [s for s in succ_states if s not in \
            visited_nodes if s not in obstacles]
```

The function is generating all the possible moves from a current field that does not end up being blocked by an obstacle. We also add a filter to exclude moves that return to a field we have visited already to avoid infinite loops.

5. Next, implement the initial costs, as shown in the following code snippet:

```
def initialize_costs(size, start):
    (h, w) = size
    costs = [[math.inf] * w for i in range(h)]
    (x, y) = start
    costs[x-1][y-1] = 0
    return costs
```

6. Now, implement the updated costs using **costs**, **current_node**, and **successor_node**:

```
def update_costs(costs, current_node, successor_nodes):
    new_cost = costs[current_node[0]-1]\
                [current_node[1]-1] + 1
    for (x, y) in successor_nodes:
        costs[x-1][y-1] = min(costs[x-1][y-1], new_cost)
```

7. Finally, implement the BFS algorithm to search the state of the tree and save the result in a variable called **bfs**:

```
def bfs_tree(node):
    nodes_to_visit = [node]
    visited_nodes = []
    costs = initialize_costs(size, start)
    while len(nodes_to_visit) > 0:
        current_node = nodes_to_visit.pop(0)
        visited_nodes.append(current_node)
        successor_nodes = successors(current_node, \
                                     visited_nodes)
        update_costs(costs, current_node, successor_nodes)
        nodes_to_visit.extend(successor_nodes)
    return costs
bfs = bfs_tree(start)
bfs
```

In the preceding code snippet, we have reused the **bfs_tree** function that we looked at earlier in the *Breadth First Search* section of this book. However, we added the **update_costs** function to update the costs.

The expected output is this:

```
[[6, 5, 4, 5, 6, 7, 8, 9, 10],
 [5, 4, 3, 4, 5, 6, 7, 8, 9],
 [4, 3, 2, inf, inf, inf, inf, inf, 10],
 [3, 2, 1, 2, inf, 12, 13, 12, 11],
 [2, 1, 0, 1, inf, 11, inf, 13, inf],
 [3, inf, inf, inf, inf, 10, inf, 14, 15],
 [4, 5, 6, 7, 8, 9, inf, 15, 16]]
```

Here, you can see that a simple BFS algorithm successfully determines the cost from the start node to any nodes, including the target node.

8. Now, measure the number of steps required to find the goal node and save the result in the **bfs_v** variable, as shown in the following code snippet:

```
def bfs_tree_verbose(node):
    nodes_to_visit = [node]
    visited_nodes = []
    costs = initialize_costs(size, start)
    step_counter = 0
    while len(nodes_to_visit) > 0:
        step_counter += 1
        current_node = nodes_to_visit.pop(0)
        visited_nodes.append(current_node)
        successor_nodes = successors(current_node, \
                                     visited_nodes)
        update_costs(costs, current_node, successor_nodes)
        nodes_to_visit.extend(successor_nodes)
        if current_node == end:
            print('End node has been reached in ', \
                  step_counter, ' steps')
            return costs
    return costs

bfs_v = bfs_tree_verbose(start)
bfs_v
```

In the preceding code snippet, we have added a step counter variable in order to print the number of steps at the end of the search.

The expected output is this:

```
End node has been reached in 110 steps
[[6, 5, 4, 5, 6, 7, 8, 9, 10],
 [5, 4, 3, 4, 5, 6, 7, 8, 9],
 [4, 3, 2, inf, inf, inf, inf, inf, 10],
 [3, 2, 1, 2, inf, 12, 13, 12, 11],
 [2, 1, 0, 1, inf, 11, inf, 13, inf],
 [3, inf, inf, inf, inf, 10, inf, 14, 15],
 [4, 5, 6, 7, 8, 9, inf, 15, 16]]
```

> **NOTE**
>
> To access the source code for this specific section, please refer to
> https://packt.live/3fMYwEt.
>
> You can also run this example online at https://packt.live/3duuLqp. You must
> execute the entire Notebook in order to get the desired result.

In this exercise, we used the BFS algorithm to find the shortest path to the goal. It took BFS **110** steps to reach the goal. Now, we will learn about an algorithm that can find the shortest path from the start node to the goal node: the A* algorithm.

INTRODUCING THE A* ALGORITHM

A* is a complete and optimal heuristic search algorithm that finds the shortest possible path between the current game state and the winning state. The definition of complete and optimal in this state are as follows:

- Complete means that A* always finds a solution.
- Optimal means that A* will find the best solution.

To set up the A* algorithm, we need the following:

- An initial state

- A description of the goal states

- Admissible heuristics to measure progress toward the goal state

- A way to generate the next steps toward the goal

Once the setup is complete, we execute the A* algorithm using the following steps on the initial state:

1. We generate all possible next steps.

2. We store these children in the order of their distance from the goal.

3. We select the child with the best score first and repeat these three steps on the child with the best score as the initial state. This is the shortest path to get to a node from the starting node.

Let's take, for example, the following figure:

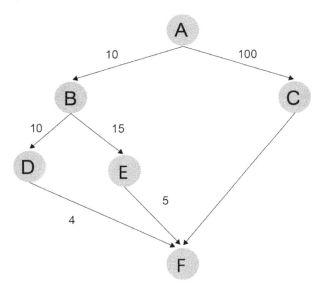

Figure 1.15: Tree with heuristic distance

The first step will be to generate all the possible moves from the origin, **A**, which is moving from **A** to **B** **(A,B)** or to **C** **(A,C)**.

The second step is to use the heuristic (the distance) to order the two possible moves, **(A,B)**, with **10**, which is shorter compared to **(A,C)** with **100**.

The third step is to choose the shortest heuristic, which is **(A,B)**, and move to **B**.

Now, we will repeat the same steps with **B** as the origin.

At the end, we will reach the goal **F** with the path (**A,B,D,F**) with a cumulative heuristic of 24. If we were following another path, such as (**A,B,E,F**), the cumulative heuristic will be 30, which is higher than the shortest path.

We did not even look at (**A,C,F**) as it was already way over the shortest path.

In pathfinding, a good heuristic is the Euclidean distance. If the current node is (x, y) and the goal node is (u, v), then we have the following:

*distance_from_end(node) = sqrt(abs(x – u) ** 2 + abs(y – v) ** 2)*

Here, **distance_from_end(node)** is an admissible heuristic estimation showing how far we are from the goal node.

We also have the following:

- **sqrt** is the square root function. Do not forget to import it from the **math** library.

- **abs** is the absolute value function, that is, **abs(-2) = abs(2) = 2**.

- **x ** 2** is **x** raised to the second power.

We will use the **distance_from_start** matrix to store the distances from the start node. In the algorithm, we will refer to this cost matrix as **distance_from_start(n1)**. For any node, **n1**, that has coordinates **(x1, y1)**, this distance is equivalent to **distance_from_start[x1][y1]**.

We will use the **succ(n)** notation to generate a list of successor nodes from **n**.

> **NOTE**
>
> The # symbol in the code snippet below denotes a code comment. Comments are added into code to help explain specific bits of logic. The triple-quotes (""") shown in the code snippet below are used to denote the start and end points of a multi-line code comment. Comments are added into code to help explain specific bits of logic.

Have a look at the pseudocode of the algorithm:

```
frontier = [start], internal = {}
# Initialize the costs matrix with each cell set to infinity.
# Set the value of distance_from_start(start) to 0.
while frontier is not empty:
    """
    notice n has the lowest estimated total
    distance between start and end.
    """
    n = frontier.pop()
    # We'll learn later how to reconstruct the shortest path
    if n == end:
        return the shortest path.
    internal.add(n)
    for successor s in succ(n):
        if s in internal:
            continue # The node was already examined
        new_distance = distance_from_start(n) + distance(n, s)
        if new_distance >= distance_from_start(s):
            """
            This path is not better than the path we have
            already examined.
            """
            continue
        if s is a member of frontier:
            update the priority of s
        else:
            Add s to frontier.
```

Regarding the retrieval of the shortest path, we can use the **costs** matrix. This matrix contains the distance of each node on the path from the start node. As cost always decreases when walking backward, all we need to do is start with the end node and walk backward greedily toward decreasing costs:

```
path = [end_node], distance = get_distance_from_start( end_node )
while the distance of the last element in the path is not 0:
    for each neighbor of the last node in path:
        new_distance = get_distance_from_start( neighbor )
```

```
        if new_distance < distance:
            add neighbor to path, and break out from the for loop
    return path
```

A* shines when we have one start state and one goal state. The complexity of the A* algorithm is **O (E)**, where **E** stands for all possible edges in the field. In our example, we have up to four edges leaving any node: up, down, left, and right.

> **NOTE**
>
> To sort the frontier list in the proper order, we must use a special Python data structure: a priority queue.

Have a look at the following example:

```
# Import heapq to access the priority queue
import heapq
# Create a list to store the data
data = []
"""
Use heapq.heappush to push (priorityInt, value)
pairs to the queue
"""
heapq.heappush(data, (2, 'first item'))
heapq.heappush(data, (1, 'second item'))
"""
The tuples are stored in data in the order
of ascending priority
"""
[(1, 'second item'), (2, 'first item')]
"""
heapq.heappop pops the item with the lowest score
from the queue
"""
heapq.heappop(data)
```

The expected output is this:

```
(1, 'second item')
```

The data still contains the second item. If you type in the following command, you will be able to see it:

```
data
```

The expected output is this:

```
[(2, 'first item')]
```

Why is it important that the heuristic being used by the algorithm is admissible?

Because this is how we guarantee the optimal nature of the algorithm. For any node **x**, we are measuring the sum of the distances from the start node to **x**. This is the estimated distance from **x** to the end node. If the estimation never overestimates the distance from **x** to the end node, we will never overestimate the total distance. Once we are at the goal node, our estimation is zero, and the total distance from the start to the end becomes an exact number.

We can be sure that our solution is optimal because there are no other items in the priority queue that have a lower estimated cost. Given that we never overestimate our costs, we can be sure that all of the nodes in the frontier of the algorithm have either similar total costs or higher total costs than the path we found.

In the following example, we can see how to implement the A* algorithm to find the path with the lowest cost:

Figure 1.16: Shortest pathfinding game board

We import **math** and **heapq**:

```
import math
import heapq
```

Next, we'll reuse the initialization code from *Steps 2–5* of the previous, *Exercise 1.05, Finding the Shortest Path Using BFS.*

> **NOTE**
>
> We have omitted the function to update costs because we will do so within the A* algorithm:

Next, we need to initialize the A* algorithm's frontier and internal lists. For **frontier**, we will use a Python **PriorityQueue**. Do not execute this code directly; we will use these four lines inside the A* search function:

```
frontier = []
internal = set()
heapq.heappush(frontier, (0, start))
costs = initialize_costs(size, start)
```

Now, it is time to implement a heuristic function that measures the distance between the current node and the goal node using the algorithm we saw in the heuristic section:

```
def distance_heuristic(node, goal):
    (x, y) = node
    (u, v) = goal
    return math.sqrt(abs(x - u) ** 2 + abs(y - v) ** 2)
```

The final step will be to translate the A* algorithm into functioning code:

```
def astar(start, end):
    frontier = []
    internal = set()
    heapq.heappush(frontier, (0, start))
    costs = initialize_costs(size, start)

    def get_distance_from_start(node):
```

```
        return costs[node[0] - 1][node[1] - 1]

    def set_distance_from_start(node, new_distance):
        costs[node[0] - 1][node[1] - 1] = new_distance


    while len(frontier) > 0:
        (priority, node) = heapq.heappop(frontier)
        if node == end:
            return priority
        internal.add(node)
        successor_nodes = successors(node, internal)
        for s in successor_nodes:
            new_distance = get_distance_from_start(node) + 1
            if new_distance < get_distance_from_start(s):
                set_distance_from_start(s, new_distance)
                # Filter previous entries of s
                frontier = [n for n in frontier if s != n[1]]
                heapq.heappush(frontier, \
                        (new_distance \
                        + distance_heuristic(s, end), s))


astar(start, end)
```

The expected output is this:

```
15.0
```

There are a few differences between our implementation and the original algorithm.

We defined a **distance_from_start** function to make it easier and more semantic to access the **costs** matrix. Note that we number the node indices starting with 1, while in the matrix, indices start with zero. Therefore, we subtract 1 from the node values to get the indices.

When generating the successor nodes, we automatically ruled out nodes that are in the internal set. **successors = succ(node, internal)** makes sure that we only get the neighbors whose examination is not closed yet, meaning that their score is not necessarily optimal.

Therefore, we may skip the step check since internal nodes will never end up in **succ(n)**.

Since we are using a priority queue, we must determine the estimated priority of nodes before inserting them. However, we will only insert the node in the frontier if we know that this node does not have an entry with a lower score.

It may happen that nodes are already in the frontier queue with a higher score. In this case, we remove this entry before inserting it into the right place in the priority queue. When we find the end node, we simply return the length of the shortest path instead of the path itself.

To follow what the A* algorithm does, execute the following example code and observe the logs:

```
def astar_verbose(start, end):
    frontier = []
    internal = set()
    heapq.heappush(frontier, (0, start))
    costs = initialize_costs(size, start)

    def get_distance_from_start(node):
        return costs[node[0] - 1][node[1] - 1]

    def set_distance_from_start(node, new_distance):
        costs[node[0] - 1][node[1] - 1] = new_distance

    steps = 0
    while len(frontier) > 0:
        steps += 1
        print('step ', steps, '. frontier: ', frontier)
        (priority, node) = heapq.heappop(frontier)
        print('node ', node, \
                'has been popped from frontier with priority', \
                priority)
        if node == end:
            print('Optimal path found. Steps: ', steps)
            print('Costs matrix: ', costs)
            return priority
        internal.add(node)
        successor_nodes = successors(node, internal)
        print('successor_nodes', successor_nodes)
        for s in successor_nodes:
            new_distance = get_distance_from_start(node) + 1
```

```
            print('s:', s, 'new distance:', new_distance, \
                ' old distance:', get_distance_from_start(s))
        if new_distance < get_distance_from_start(s):
            set_distance_from_start(s, new_distance)
            # Filter previous entries of s
            frontier = [n for n in frontier if s != n[1]]
            new_priority = new_distance \
                            + distance_heuristic(s, end)
            heapq.heappush(frontier, (new_priority, s))
            print('Node', s, \
                'has been pushed to frontier with priority', \
                new_priority)
    print('Frontier', frontier)
    print('Internal', internal)
    print(costs)
astar_verbose(start, end)
```

Here, we build the **astar_verbose** function by reusing the code from the **astar** function and adding print functions in order to create a log.

The expected output is this:

```
node (1, 9) has been popped from frontier with priority 15.0
successor_nodes []
step 42 . frontier: [(15.0, (6, 8)), (15.60555127546399, (4, 6)), (15.433981132056603, (1, 1)), (15.8284271247461
9, (4, 7))]
node (6, 8) has been popped from frontier with priority 15.0
successor_nodes [(7, 8), (6, 9)]
s: (7, 8) new distance: 15  old distance: inf
Node (7, 8) has been pushed to frontier with priority 16.414213562373096
s: (6, 9) new distance: 15  old distance: inf
Node (6, 9) has been pushed to frontier with priority 15.0
step 43 . frontier: [(15.0, (6, 9)), (15.433981132056603, (1, 1)), (15.82842712474619, (4, 7)), (16.4142135623730
96, (7, 8)), (15.60555127546399, (4, 6))]
node (6, 9) has been popped from frontier with priority 15.0
Optimal path found. Steps: 43
Costs matrix: [[6, 5, 4, 5, 6, 7, 8, 9, 10], [5, 4, 3, 4, 5, 6, 7, 8, 9], [4, 3, 2, inf, inf, inf, inf, inf, 10],
[3, 2, 1, 2, inf, 12, 13, 12, 11], [2, 1, 0, 1, inf, 11, inf, 13, inf], [3, inf, inf, inf, inf, 10, inf, 14, 15],
[4, 5, 6, 7, 8, 9, inf, 15, inf]]
Out[7]: 15.0
```

Figure 1.17: Astar function logs

We have seen that the A* search returns the right values. The question is, how can we reconstruct the whole path?

For this, we remove the **print** statements from the code for clarity and continue with the A* algorithm that we implemented in the previous step. Instead of returning the length of the shortest path, we have to return the path itself. We will write a function that extracts this path by walking backward from the end node, analyzing the **costs** matrix. Do not define this function globally yet. We define it as a local function in the A* algorithm that we created previously:

```python
def get_shortest_path(end_node):
    path = [end_node]
    distance = get_distance_from_start(end_node)
    while distance > 0:
        for neighbor in successors(path[-1], []):
            new_distance = get_distance_from_start(neighbor)
            if new_distance < distance:
                path += [neighbor]
                distance = new_distance
                break  # for
    return path
```

Now that we've seen how to deconstruct the path, let's return it inside the A* algorithm:

```python
def astar_with_path(start, end):
    frontier = []
    internal = set()
    heapq.heappush(frontier, (0, start))
    costs = initialize_costs(size, start)

    def get_distance_from_start(node):
        return costs[node[0] - 1][node[1] - 1]

    def set_distance_from_start(node, new_distance):
        costs[node[0] - 1][node[1] - 1] = new_distance

    def get_shortest_path(end_node):
        path = [end_node]
        distance = get_distance_from_start(end_node)
        while distance > 0:
            for neighbor in successors(path[-1], []):
                new_distance = get_distance_from_start(neighbor)
                if new_distance < distance:
```

```
                    path += [neighbor]
                    distance = new_distance
                    break   # for
        return path


    while len(frontier) > 0:
        (priority, node) = heapq.heappop(frontier)
        if node == end:
            return get_shortest_path(end)
        internal.add(node)
        successor_nodes = successors(node, internal)
        for s in successor_nodes:
            new_distance = get_distance_from_start(node) + 1
            if new_distance < get_distance_from_start(s):
                set_distance_from_start(s, new_distance)
                # Filter previous entries of s
                frontier = [n for n in frontier if s != n[1]]
                heapq.heappush(frontier, \
                            (new_distance \
                            + distance_heuristic(s, end), s))


astar_with_path( start, end )
```

In the preceding code snippet, we have reused the **a-star** function defined previously with the notable difference of adding the **get_shortest_path** function. Then, we use this function to replace the priority queue since we want the algorithm to always choose the shortest path.

The expected output is this:

```
Out[9]:  [(6, 9),
          (6, 8),
          (5, 8),
          (4, 8),
          (4, 9),
          (3, 9),
          (2, 9),
          (2, 8),
          (2, 7),
          (2, 6),
          (2, 5),
          (2, 4),
          (2, 3),
          (3, 3),
          (4, 3),
          (5, 3)]
```

Figure 1.18: Output showing the priority queue

Technically, we do not need to reconstruct the path from the **costs** matrix. We could record the parent node of each node in the matrix and simply retrieve the coordinates to save a bit of searching.

We are not expecting you to understand all the preceding script as it is quite advanced, so we are going to use a library that will simplify it for us.

A* SEARCH IN PRACTICE USING THE SIMPLEAI LIBRARY

The **simpleai** library is available on GitHub and contains many popular AI tools and techniques.

> **NOTE**
>
> You can access this library at https://github.com/simpleai-team/simpleai. The documentation of the **simpleai** library can be accessed here: http://simpleai.readthedocs.io/en/latest/. To access the **simpleai** library, first, you have to install it.

The **simpleai** library can be installed as follows:

```
pip install simpleai
```

Once **simpleai** has been installed, you can import classes and functions from the **simpleai** library into a Jupyter Notebook:

```
from simpleai.search import SearchProblem, astar
```

SearchProblem gives you a frame for defining any search problems. The **astar** import is responsible for executing the A* algorithm inside the search problem.

For simplicity, we have not used classes in the previous code examples to focus on the algorithms in a plain old style without any clutter.

> **NOTE**
>
> Remember that the **simpleai** library will force us to use classes.

To describe a search problem, you need to provide the following:

- **constructor**: This initializes the state space, thus describing the problem. We will make the **Size**, **Start**, **End**, and **Obstacles** values available in the object by adding it to these as properties. At the end of the constructor, do not forget to call the super constructor, and do not forget to supply the initial state.

- **actions(state)**: This returns a list of actions that we can perform from a given state. We will use this function to generate new states. Semantically, it would make more sense to create action constants such as UP, DOWN, LEFT, and RIGHT, and then interpret these action constants as a result. However, in this implementation, we will simply interpret an action as "move to **(x, y)**", and represent this command as **(x, y)**. This function contains more-or-less the logic that we implemented in the **succ** function previously, except that we won't filter the result based on a set of visited nodes.

- **result(state0, action)**: This returns the new state of action that was applied to **state0**.

- **is_goal(state)**: This returns **true** if the state is a goal state. In our implementation, we will have to compare the state to the end state coordinates.

- **cost(self, state, action, newState)**: This is the cost of moving from **state** to **newState** via **action**. In our example, the cost of a move is uniformly 1.

Have a look at the following example:

```python
import math
from simpleai.search import SearchProblem, astar

class ShortestPath(SearchProblem):
    def __init__(self, size, start, end, obstacles):
        self.size = size
        self.start = start
        self.end = end
        self.obstacles = obstacles
        super(ShortestPath, \
              self).__init__(initial_state=self.start)

    def actions(self, state):
        (row, col) = state
        (max_row, max_col) = self.size
        succ_states = []
        if row > 1:
            succ_states += [(row-1, col)]
        if col > 1:
            succ_states += [(row, col-1)]
        if row < max_row:
```

```
                succ_states += [(row+1, col)]
            if col < max_col:
                succ_states += [(row, col+1)]
            return [s for s in succ_states \
                    if s not in self.obstacles]

    def result(self, state, action):
        return action

    def is_goal(self, state):
        return state == end

    def cost(self, state, action, new_state):
        return 1

    def heuristic(self, state):
        (x, y) = state
        (u, v) = self.end
        return math.sqrt(abs(x-u) ** 2 + abs(y-v) ** 2)

size = (7, 9)
start = (5, 3)
end = (6, 9)
obstacles = {(3, 4), (3, 5), (3, 6), (3, 7), (3, 8), \
             (4, 5), (5, 5), (5, 7), (5, 9), (6, 2), \
             (6, 3), (6, 4), (6, 5), (6, 7), (7, 7)}
searchProblem = ShortestPath(size, start, end, obstacles)

result = astar(searchProblem, graph_search=True)
result.path()
```

In the preceding code snippet, we used the **simpleai** package to simplify our code. We also had to define a class called **ShortestPath** in order to use the package.

The expected output is this:

```
[(None, (5, 3)),
 ((4, 3), (4, 3)),
 ((3, 3), (3, 3)),
 ((2, 3), (2, 3)),
 ((2, 4), (2, 4)),
 ((2, 5), (2, 5)),
 ((2, 6), (2, 6)),
 ((2, 7), (2, 7)),
 ((2, 8), (2, 8)),
 ((2, 9), (2, 9)),
 ((3, 9), (3, 9)),
 ((4, 9), (4, 9)),
 ((4, 8), (4, 8)),
 ((5, 8), (5, 8)),
 ((6, 8), (6, 8)),
 ((6, 9), (6, 9))]
```

Figure 1.19: Output showing the queue using the simpleai library

The **simpleai** library made the search description a lot easier than the manual implementation. All we need to do is define a few basic methods, and then we have access to an effective search implementation.

In the next section, we will be looking at the Minmax algorithm, along with pruning.

GAME AI WITH THE MINMAX ALGORITHM AND ALPHA-BETA PRUNING

In the first two sections, we saw how hard it was to create a winning strategy for a simple game such as tic-tac-toe. The previous section introduced a few structures for solving search problems with the A* algorithm. We also saw that tools such as the **simpleai** library help us to reduce the effort we put in to describe a task with code.

We will use all of this knowledge to supercharge our game AI skills and solve more complex problems.

SEARCH ALGORITHMS FOR TURN-BASED MULTIPLAYER GAMES

Turn-based multiplayer games such as tic-tac-toe are similar to pathfinding problems. We have an initial state and we have a set of end states where we win the game.

The challenge with turn-based multiplayer games is the combinatorial explosion of the opponent's possible moves. This difference justifies treating turn-based games differently to a regular pathfinding problem.

For instance, in the tic-tac-toe game, from an empty board, we can select one of the nine cells and place our sign there, assuming we start the game. Let's denote this algorithm with the **succ** function, symbolizing the creation of successor states. Consider we have the initial state denoted by **Si**.

Here, we have **succ(Si) returns [S1, S2, ..., Sn]**, where **S1, S2, ..., Sn** are successor states:

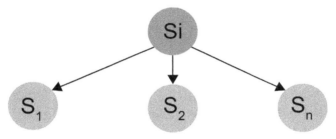

Figure 1.20: Tree diagram denoting the successor states of the function

Then, the opponent also makes a move, meaning that from each possible state, we have to examine even more states:

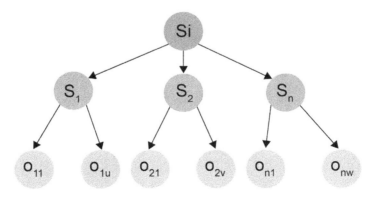

Figure 1.21: Tree diagram denoting parent-successor relationships

The expansion of possible future states stops in one of two cases:

- The game ends.

- Due to resource limitations, it is not worth explaining any more moves beyond a certain depth for the state of a certain utility.

Once we stop expanding, we have to make a static heuristic evaluation of the state. This is exactly what we did previously with the A* algorithm, when choosing the best move; however, we never considered future states.

Therefore, even though our algorithm became more and more complex, without using the knowledge of possible future states, we had a hard time detecting whether our current move would likely be a winning one or a losing one.

The only way for us to take control of the future was to change our heuristic while knowing how many games we would win, lose, or tie in the future. We could either maximize our wins or minimize our losses. We still did not dig deep enough to see whether our losses could have been avoided through smarter play on the part of the AI.

All these problems can be avoided by digging deeper into future states and recursively evaluating the utility of the branches.

To consider future states, we will learn about the **Minmax** algorithm and its variant, the **NegaMax** algorithm.

THE MINMAX ALGORITHM

Suppose there is a game where a heuristic function can evaluate a game state from the perspective of the AI player. For instance, we used a specific evaluation for the tic-tac-toe exercise:

- +1,000 points for a move that won the game

- +100 points for a move preventing the opponent from winning the game

- +10 points for a move creating two in a row, column, or diagonal

- +1 point for a move creating one in a row, column, or diagonal

This static evaluation is straightforward to implement on any node. The problem is that as we go deep into the tree of all possible future states, we do not yet know what to do with these scores. This is where the Minmax algorithm comes into play.

Suppose we construct a tree with each possible move that could be performed by each player up to a certain depth. At the bottom of the tree, we evaluate each option. For the sake of simplicity, let's assume that we have a search tree that appears as follows:

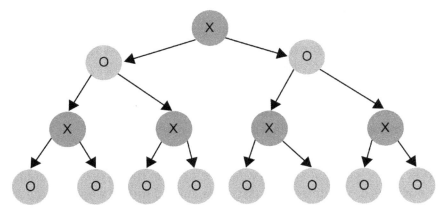

Figure 1.22: Example of a search tree up to a certain depth

The AI plays with **X**, and the player plays with **O**. A node with **X** means that it is **X**'s turn to move. A node with **O** means it is **O**'s turn to act.

Suppose there are all **O** leaves at the bottom of the tree, and we didn't compute any more values because of resource limitations. Our task is to evaluate the utility of the leaves:

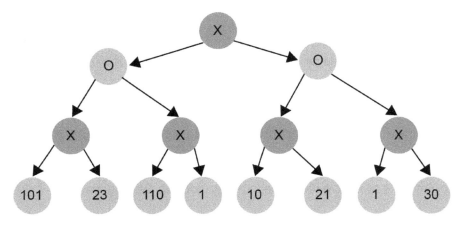

Figure 1.23: Example of a search tree with possible moves

We have to select the best possible move from our perspective because our goal is to maximize the utility of our move. This aspiration to maximize our gains represents the Max part in the Minmax algorithm:

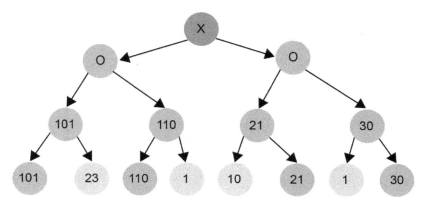

Figure 1.24: Example of a search tree with the best possible move

If we move one level higher, it is our opponent's turn to act. Our opponent picks the value that is the least beneficial to us. This is because our opponent's job is to minimize our chances of winning the game. This is the Min part of the Minmax algorithm:

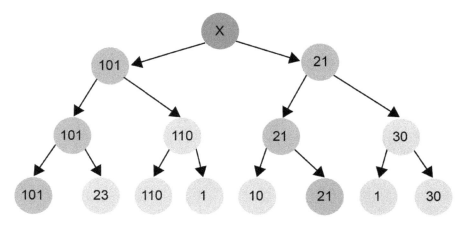

Figure 1.25: Minimizing the chances of winning the game

At the top, we can choose between a move with utility **101** and another move with utility **21**. Since we are maximizing our value, we should pick **101**:

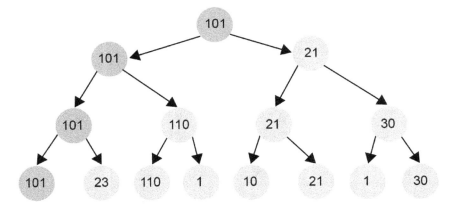

Figure 1.26: Maximizing the chances of winning the game

Let's see how we can implement this idea:

```
def min_max( state, depth, is_maximizing):
    if depth == 0 or is_end_state( state ):
        return utility( state )
    if is_maximizing:
        utility = 0
        for s in successors( state ):
            score = MinMax( s, depth - 1, false )
            utility = max( utility, score )
        return utility
    else:
        utility = infinity
        for s in successors( state ):
            score = MinMax( s, depth - 1, true )
            utility = min( utility, score )
        return utility
```

This is the Minmax algorithm. We evaluate the leaves from our perspective. Then, from the bottom up, we apply a recursive definition:

- Our opponent plays optimally by selecting the worst possible node from our perspective.

- We play optimally by selecting the best possible node from our perspective.

We need a few more things in order to understand the application of the Minmax algorithm on the tic-tac-toe game:

- **is_end_state** is a function that determines whether the state should be evaluated instead of digging deeper, either because the game has ended, or because the game is about to end using forced moves. Using our utility function, it is safe to say that as soon as we reach a score of 1,000 or higher, we have effectively won the game. Therefore, **is_end_state** can simply check the score of a node and determine whether we need to dig deeper.

- Although the **successors** function only depends on the state, it is practical to pass the information of whose turn it is to make a move. Therefore, do not hesitate to add an argument if needed; you do not have to follow the pseudo code.

- We want to minimize our efforts in implementing the Minmax algorithm. For this reason, we will evaluate existing implementations of the algorithm. We will also simplify the duality of the description of the algorithm in the remainder of this section.

- The suggested utility function is quite accurate compared to the utility functions that we could be using in this algorithm. In general, the deeper we go, the less accurate our utility function has to be. For instance, if we could go nine steps deep into the tic-tac-toe game, all we would need to do is award 1 point for a win, 0 for a draw, and -1 point for a loss, given that, in nine steps, the board is complete, and we have all of the necessary information to make the evaluation. If we could only look four steps deep, this utility function would be completely useless at the start of the game because we need at least five steps to win the game.

- The Minmax algorithm could be optimized further by pruning the tree. Pruning is an act where we get rid of branches that do not contribute to the result. By eliminating unnecessary computations, we save precious resources that could be used to go deeper into the tree.

OPTIMIZING THE MINMAX ALGORITHM WITH ALPHA-BETA PRUNING

The last consideration in the previous thought process primed us to explore possible optimizations by reducing the search space by focusing our attention on nodes that matter.

There are a few constellations of nodes in the tree where we can be sure that the evaluation of a subtree does not contribute to the end result. We will find, examine, and generalize these constellations to optimize the Minmax algorithm.

Let's examine pruning through the previous example of nodes:

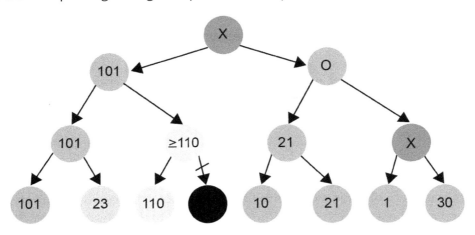

Figure 1.27: Search tree demonstrating pruning nodes

After computing the nodes with values **101**, **23**, and **110**, we can conclude that two levels above, the value **101** will be chosen. Why?

- Suppose X <= 110. Here, the maximum of **110** and X will be chosen, which is **110**, and X will be omitted.

- Suppose X > 110. Here, the maximum of **110** and X is X. One level above, the algorithm will choose the lowest value out of the two. The minimum of **101** and X will always be **101**, because X > 110. Therefore, X will be omitted a level above.

This is how we prune the tree.

On the right-hand side, suppose we computed branches **10** and **21**. Their maximum is **21**. The implication of computing these values is that we can omit the computation of nodes Y1, Y2, and Y3, and we know that the value of Y4 is less than or equal to **21**. Why?

The minimum of **21** and Y3 is never greater than **21**. Therefore, Y4 will never be greater than **21**.

We can now choose between a node with utility **101** and another node with a maximal utility of **21**. It is obvious that we have to choose the node with utility **101**:

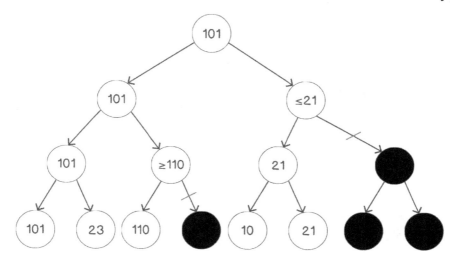

Figure 1.28: Example of pruning a tree

This is the idea behind alpha-beta pruning. We prune subtrees that we know are not going to be needed.

Let's see how we can implement alpha-beta pruning in the Minmax algorithm.

First, we will add an alpha and a beta argument to the argument list of Minmax:

```
def min_max(state, depth, is_maximizing, alpha, beta):
    if depth == 0 or is_end_state(state):
        return utility(state)
    if is_maximizing:
        utility = 0
        for s in successors(state):
            score = MinMax(s, depth - 1, false, alpha, beta)
            utility = max(utility, score)
            alpha = max(alpha, score)
            if beta <= alpha:
                break
        return utility
    else:
        utility = infinity
```

```
    for s in successors(state):
        score = MinMax(s, depth - 1, true, alpha, beta)
        utility = min(utility, score)
    return utility
```

In the preceding code snippet, we added the **alpha** and **beta** arguments to the **MinMax** function in order to calculate the new alpha score as being the maximum between **alpha** and **beta** in the maximizing branch.

Now, we need to do the same with the minimizing branch:

```
def min_max(state, depth, is_maximizing, alpha, beta):
    if depth == 0 or is_end_state( state ):
        return utility(state)
    if is_maximizing:
        utility = 0
        for s in successors(state):
            score = min_max(s, depth - 1, false, alpha, beta)
            utility = max(utility, score)
            alpha = max(alpha, score)
            if beta <= alpha: break
        return utility
    else:
        utility = infinity
        for s in successors(state):
            score = min_max(s, depth - 1, true, alpha, beta)
            utility = min(utility, score)
            beta = min(beta, score)
            if beta <= alpha: break
        return utility
```

In the preceding code snippet, we added the new beta score in the **else** branch, which is the minimum between **alpha** and **beta** in the minimizing branch.

We are done with the implementation. It is recommended that you mentally execute the algorithm on our example tree step by step to get a feel for the implementation.

One important piece is missing that has prevented us from doing the execution properly: the initial values for **alpha** and **beta**. Any number that is outside the possible range of utility values will do. We will use positive and negative infinity as initial values to call the Minmax algorithm:

```
alpha = infinity
beta = -infinity
```

In the next section, we will look at the DRYing technique while using the NegaMax algorithm.

DRYING UP THE MINMAX ALGORITHM – THE NEGAMAX ALGORITHM

The Minmax algorithm works great, especially with alpha-beta pruning. The only problem is that we have **if** and **else** branches in the algorithm that essentially negates each other.

As we know, in computer science, there is DRY code and WET code. **DRY** stands for **Don't Repeat Yourself**. **WET** stands for **Write Everything Twice**. When we write the same code twice, we double our chance of making a mistake while writing it. We also double our chances of each maintenance effort being executed in the future. Hence, it is better to reuse our code.

When implementing the Minmax algorithm, we always compute the utility of a node from the perspective of the AI player. This is why we have to have a utility-maximizing branch and a utility-minimizing branch in the implementations that are dual in nature. As we prefer clean code that describes the problem only once, we could get rid of this duality by changing the point of view of the evaluation.

Whenever the AI player's turn comes, nothing changes in the algorithm.

Whenever the opponent's turn comes, we negate the perspective. Minimizing the AI player's utility is equivalent to maximizing the opponent's utility.

This simplifies the Minmax algorithm:

```
def Negamax(state, depth, is_players_point_of_view):
    if depth == 0 or is_end_state(state):
        return utility(state, is_players_point_of_view)
    utility = 0
    for s in successors(state):
        score = Negamax(s,depth-1,not is_players_point_of_view)
    return score
```

There are necessary conditions for using the NegaMax algorithm; for instance, the evaluation of the board state has to be symmetric. If a game state is worth +20 from the first player's perspective, it is worth -20 from the second player's perspective. Therefore, we often normalize the scores around zero.

USING THE EASYAI LIBRARY

We have already looked at the **simpleai** library, which helped us execute searches on pathfinding problems. Now, we will use the **EasyAI** library, which can easily handle an AI search on two-player games, reducing the implementation of the tic-tac-toe problem to writing a few functions on scoring the utility of a board and determining when the game ends.

To install **EasyAI**, type the following command in Jupyter Notebook:

```
!pip install easyAI
```

> **NOTE**
>
> You can read the documentation of the library on GitHub at https://github.com/Zulko/easyAI.

ACTIVITY 1.04: CONNECT FOUR

In this activity, we will practice using the **EasyAI** library and develop a heuristic. We will be using the game *Connect Four* for this. The game board is seven cells wide and seven cells high. When you make a move, you can only select the column in which you drop your token. Then, gravity pulls the token down to the lowest possible empty cell. Your objective is to connect four of your own tokens horizontally, vertically, or diagonally, before your opponent does, or you run out of empty spaces.

> **NOTE**
>
> The rules of the game can be found at
> https://en.wikipedia.org/wiki/Connect_Four.

1. Open a new Jupyter Notebook file.

2. Write the **init** method to generate all the possible winning combinations in the game and save them for future use.

3. Write a function to enumerate all the possible moves. Then, for each column, check whether there is an unoccupied field. If there is one, make the column a possible move.

4. Create a function to make a move (it will be similar to the possible move function), and then check the column of the move and find the first empty cell, starting from the bottom.

5. Reuse the lose function from the tic-tac-toe example.

6. Implement the show method that prints the board and try out the game.

> **NOTE**
>
> The solution to this activity can be found on page 330.

The expected output is this:

```
Player 1 what do you play ? 5

Move #9: player 1 plays 5 :

. . . . . . .
. . . . . . .
X . . . . . .
X . . . . . .
X . . O . . .
O O O X O . .

Move #10: player 2 plays 1 :

. . . . . . .
X . . . . . .
X . . . . . .
X . . . . . .
X . . O . . .
O O O X O . .
```

Figure 1.29: Expected output for the game Connect Four

SUMMARY

In this chapter, we have seen how AI can be used to enhance or substitute human abilities such as to listen, speak, understand language, store and retrieve information, think, see, and move.

Then, we moved on to learning about intelligent agents and the way they interact with the environment, solving a problem in a seemingly intelligent way to pursue a goal.

Then, we introduced Python and learned about its role in AI. We looked at a few important Python libraries for developing AI and prepared data for the intelligent agents. We then created a tic-tac-toe game based on predefined rules. We quantified these rules into a number, a process that we call heuristics. We learned how to use heuristics in the A* search algorithm to find an optimal solution to a problem.

Finally, we got to know about the Minmax and NegaMax algorithms so that the AI could win two-player games. In the next chapter, you will be introduced to regression.

2

AN INTRODUCTION TO REGRESSION

OVERVIEW

In this chapter, you will be introduced to regression. Regression comes in handy when you are trying to predict future variables using historical data. You will learn various regression techniques such as linear regression with single and multiple variables, along with polynomial and Support Vector Regression (SVR). You will use these techniques to predict future stock prices from a stock price data. By the end of this chapter, you will be comfortable using regression techniques to solve practical problems in a variety of fields.

INTRODUCTION

In the previous chapter, you were introduced to the fundamentals of **Artificial Intelligence** (**AI**), which helped you create the game Tic-Tac-Toe. In this chapter, we will be looking at regression, which is a machine learning algorithm that can be used to measure how closely related independent variable(s), called **features**, relate to a dependent variable called a **label**.

Linear regression is a concept with many applications a variety of fields, ranging from finance (predicting the price of an asset) to business (predicting the sales of a product) and even the economy (predicting economy growth).

Most of this chapter will deal with different forms of linear regression, including linear regression with one variable, linear regression with multiple variables, polynomial regression with one variable, and polynomial regression with multiple variables. Python provides lots of forms of support for performing regression operations and we will also be looking at these later on in this chapter.

We will also use an alternative regression model, called **Support Vector Regression** (**SVR**), with different forms of linear regression. Throughout this chapter, we will be using a few sample datasets along with the stock price data loaded from the **Quandl** Python library to predict future prices using different types of regression.

> **NOTE**
>
> Although it is not recommended that you use the models in this chapter to provide trading or investment advice, this is a very exciting and interesting journey that explains the fundamentals of regression.

LINEAR REGRESSION WITH ONE VARIABLE

A general regression problem can be defined with the following example. Suppose we have a set of data points and we need to figure out the best fit curve to approximately fit the given data points. This curve will describe the relationship between our input variable, **x**, which is the data point, and the output variable, **y**, which is the curve.

Remember, in real life, we often have more than one input variable determining the output variable. However, linear regression with one variable will help us to understand how the input variable impacts the output variable.

TYPES OF REGRESSION

In this chapter, we will work with regression on the two-dimensional plane. This means that our data points are two-dimensional, and we are looking for a curve to approximate how to calculate one variable from another.

We will come across the following types of regression in this chapter:

- **Linear regression with one variable using a polynomial of degree 1**: This is the most basic form of regression, where a straight line approximates the trajectory of future data.

- **Linear regression with multiple variables using a polynomial of degree 1**: We will be using equations of degree 1, but we will also allow multiple input variables, called features.

- **Polynomial regression with one variable**: This is a generic form of the linear regression of one variable. As the polynomial used to approximate the relationship between the input and the output is of an arbitrary degree, we can create curves that fit the data points better than a straight line. The regression is still linear – not because the polynomial is linear, but because the regression problem can be modeled using linear algebra.

- **Polynomial regression with multiple variables**: This is the most generic regression problem, using higher degree polynomials and multiple features to predict the future.

- **SVR**: This form of regression uses **Support Vector Machines** (**SVMs**) to predict data points. This type of regression is included to explain SVR's usage compared to the other four regression types.

Now we will deal with the first type of linear regression: we will use one variable, and the polynomial of the regression will describe a straight line.

On the two-dimensional plane, we will use the Déscartes coordinate system, more commonly known as the *Cartesian coordinate system*. We have an x and a y-axis, and the intersection of these two axes is the origin. We denote points by their x and y coordinates.

For instance, point *(2, 1)* corresponds to the black point on the following coordinate system:

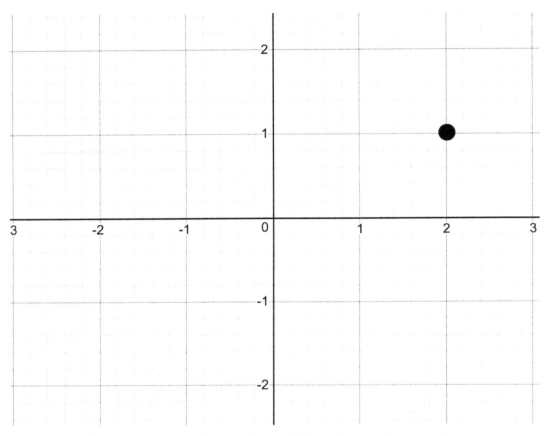

Figure 2.1: Representation of point (2,1) on the coordinate system

A straight line can be described with the equation $y = a*x + b$, where a is the slope of the equation, determining how steeply the equation climbs up, and b is a constant determining where the line intersects the y-axis.

In *Figure 2.2*, you can see three equations:

- The straight line is described with the equation $y = 2*x + 1$.

- The dashed line is described with the equation $y = x + 1$.

- The dotted line is described with the equation $y = 0.5*x + 1$.

You can see that all three equations intersect the *y*-axis at *1*, and their slope is determined by the factor by which we multiply *x*.

If you know *x*, you can solve *y*. Similarly, if you know *y*, you can solve *x*. This equation is a polynomial equation of degree *1*, which is the base of linear regression with one variable:

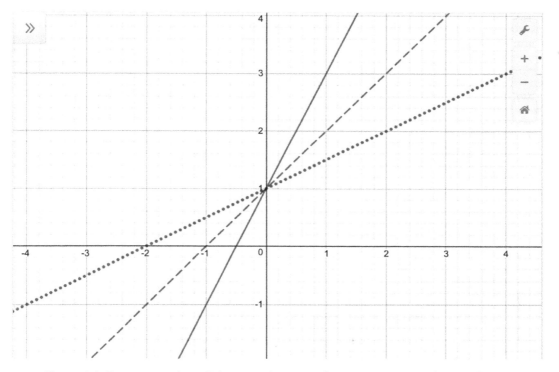

Figure 2.2: Representation of the equations y = 2*x + 1, y = x + 1, and y = 0.5*x + 1 on the coordinate system

We can describe curves instead of straight lines using polynomial equations; for example, the polynomial equation $4x^4$-$3x^3$-x^2-$3x$+3 will result in *Figure 2.3*. This type of equation is the base of polynomial regression with one variable:

Figure 2.3: Representation of the polynomial equation

> **NOTE**
>
> If you would like to experiment further with the Cartesian coordinate system, you can use the following plotter: https://s3-us-west-2.amazonaws.com/oerfiles/College+Algebra/calculator.html.

FEATURES AND LABELS

In machine learning, we differentiate between features and labels. Features are considered our **input** variables, and labels are our **output** variables.

When talking about regression, the possible value of the labels is a continuous set of rational numbers. Think of features as the values on the *x*-axis and labels as the values on the *y*-axis.

The task of regression is to predict label values based on feature values.

We often create a label by projecting the values of a feature in the future.

For instance, if we would like to predict the price of a stock for next month using historical monthly data, we would create the label by shifting the stock price feature one month into the future:

- For each stock price feature, the label would be the stock price feature of the next month.

- For the last month, prediction data would not be available, so these values are all **NaN** (Not a Number).

Let's say we have data for the months of **January**, **February**, and **March**, and we want to predict the price for **April**. Our feature for each month will be the current monthly price and the label will be the price of the next month.

For instance, take a look at the following table:

Month	Current month price (feature)	Next month price (label)
January	100	150
February	150	200
March	200	NaN
April	NaN	NaN

Figure 2.4: Example of a feature and a label

This means that the label for **January** is the price of **February** and that the label for **February** is actually the price of **March**. The label for **March** is unknown (**NaN**) as this is the value we are trying to predict.

FEATURE SCALING

At times, we have multiple features (inputs) that may have values within completely different ranges. Imagine comparing micrometers on a map to kilometers in the real world. They won't be easy to handle because of the difference in magnitude of nine zeros.

A less dramatic difference is the difference between imperial and metric data. For instance, pounds and kilograms, and centimeters and inches, do not compare that well.

Therefore, we often scale our features to normalized values that are easier to handle, as we can compare the values of these ranges more easily.

We will demonstrate two types of scaling:

- Min-max normalization

- Mean normalization

Min-max normalization is calculated as follows:

$$X_{scaled} = \frac{X - X_{MIN}}{X_{MAX} - X_{MIN}}$$

Here, X_{MIN} is the minimum value of the feature and X_{MAX} is the maximum value.

The feature-scaled values will be within the range of `[0;1]`.

Mean normalization is calculated as follows:

$$X_{scaled} = \frac{X - AVG}{X_{MAX} - X_{MIN}}$$

Here, **AVG** is the average.

The feature-scaled values will be within the range of `[-1;1]`.

Here's an example of both normalizations applied on the first 13 numbers of the Fibonacci sequence.

We begin with finding the min-max normalization:

```
fibonacci = [0, 1, 1, 2, 3, 5, 8, 13, 21, 34, 55, 89, 144]
# Min-Max normalization:
[(float(i)-min(fibonacci))/(max(fibonacci)-min(fibonacci)) \
for i in fibonacci]
```

The expected output is this:

```
[0.0,
 0.006944444444444444,
 0.006944444444444444,
 0.013888888888888888,
 0.020833333333333332,
 0.034722222222222224,
 0.05555555555555555,
 0.09027777777777778,
 0.14583333333333334,
```

```
0.2361111111111111,
0.3819444444444444,
0.6180555555555556,
1.0]
```

Now, take a look at the following code snippet to find the mean normalization:

```
# Mean normalization:
avg = sum(fibonacci) / len(fibonacci)
# 28.923076923076923
[(float(i)-avg)/(max(fibonacci)-min(fibonacci)) \
for i in fibonacci]
```

The expected output is this:

```
[-0.20085470085470086,
 -0.19391025641025642,
 -0.19391025641025642,
 -0.18696581196581197,
 -0.18002136752136752,
 -0.16613247863247863,
 -0.1452991452991453,
 -0.11057692307692307,
 -0.05502136752136752,
 0.035256410256410256,
 0.18108974358974358,
 0.4172008547008547,
 0.7991452991452992]
```

> **NOTE**
>
> Scaling could add to the processing time, but, often, it is an important step to add.

In the scikit-learn library, we have access to the **preprocessing.scale** function, which scales NumPy arrays:

```
import numpy as np
from sklearn import preprocessing
preprocessing.scale(fibonacci)
```

The expected output is this:

```
array([-0.6925069 , -0.66856384, -0.66856384, -0.64462079,
       -0.62067773-0.57279161, -0.50096244, -0.38124715,
       -0.18970269,  0.12155706, 0.62436127,  1.43842524,
       2.75529341]
```

The **scale** method performs a standardization, which is another type of normalization. Notice that the result is a NumPy array.

SPLITTING DATA INTO TRAINING AND TESTING

Now that we have learned how to normalize our dataset, we need to learn about the training-testing split. In order to measure how well our model can generalize its predictive performance, we need to split our dataset into a training set and a testing set. The training set is used by the model to learn from so that it can build predictions. Then, the model will use the testing set to evaluate the performance of its prediction.

When we split the dataset, we first need to shuffle it to ensure that our testing set will be a generic representation of our dataset. The split is usually 90% for the training set and 10% for the testing set.

With training and testing, we can measure whether our model is overfitting or underfitting.

Overfitting occurs when the trained model fits the training dataset too well. The model will be very accurate on the training data, but it will not be usable in real life, as its accuracy will decrease when used on any other data. The model adjusts to the random noise in the training data and assumes patterns on this noise that yield false predictions.

Underfitting occurs when the trained model does not fit the training data well enough to recognize important patterns in the data. As a result, it cannot make accurate predictions on new data. One example of this is when we attempt to do linear regression on a dataset that is not linear. For example, the Fibonacci sequence is not linear; therefore, a model on a Fibonacci-like sequence cannot be linear either.

We can do the training-testing split using the **model_selection** library of scikit- learn.

Suppose, in our example, that we have scaled the Fibonacci data and defined its indices as labels:

```
features = preprocessing.scale(fibonacci)
label = np.array(range(13))
```

Now, let's use 10% of the data as test data, **test_size=0.1**, and specify **random_state** parameter in order to get the exact same split every time we run the code:

```
from sklearn import model_selection
(x_train, x_test, y_train, y_test) = \
model_selection.train_test_split(features, \
                                 label, test_size=0.1, \
                                 random_state=8)
```

Our dataset has been split into test and training sets for our features (**x_train** and **x_test**) and for our labels (**y_train** and **y_test**).

Finally, let's check each set, beginning with the **x_train** feature:

```
x_train
```

The expected output is this:

```
array([ 1.43842524, -0.18970269, -0.50096244,  2.75529341,
        -0.6925069 , -0.66856384, -0.57279161,  0.12155706,
        -0.66856384, -0.62067773, -0.64462079])
```

Next, we check for **x_test**:

```
x_test
```

The expected output is this:

```
array([-0.38124715,  0.62436127])
```

Then, we check for **y_train**:

```
y_train
```

The expected output is this:

```
array([11,  8,  6, 12,  0,  2,  5,  9,  1,  4,  3])
```

Next, we check for **y_test**:

```
y_test
```

The expected output is this:

```
array([7, 10])
```

In the preceding output, we can see that our split has been properly executed; for instance, our label has been split into **y_test**, which contains the **7** and **10** indexes, and **y_train** which contains the remaining **11** indexes. The same logic has been applied to our features and we have **2** values in **x_test** and **11** values in **x_train**.

> **NOTE**
>
> If you remember the Cartesian coordinate system, you know that the horizontal axis is the *x*-axis and that the vertical axis is the *y*-axis. Our features are on the *x*-axis, while our labels are on the *y*-axis. Therefore, we use features and *x* as synonyms, while labels are often denoted by *y*. Therefore, **x_test** denotes feature test data, **x_train** denotes feature training data, **y_test** denotes label test data, and **y_train** denotes label training data.

FITTING A MODEL ON DATA WITH SCIKIT-LEARN

We are now going to illustrate the process of regression on an example where we only have one feature and minimal data.

As we only have one feature, we have to format **x_train** by reshaping it with **x_train.reshape (-1,1)** to a NumPy array containing one feature.

Therefore, before executing the code on fitting the best line, execute the following code:

```
x_train = x_train.reshape(-1, 1)
x_test = x_test.reshape(-1, 1)
```

We can fit a linear regression model on our data with the following code:

```
from sklearn import linear_model
linear_regression = linear_model.LinearRegression()
model = linear_regression.fit(x_train, y_train)
model.predict(x_test)
```

The expected output is this:

```
array([4.46396931, 7.49212796])
```

We can also calculate the score associated with the model:

```
model.score(x_test, y_test)
```

The expected output is this:

```
-1.8268608450379087
```

This score represents the accuracy of the model and is defined as the R^2 or **coefficient of determination**. It represents how well we can predict the features from the labels.

In our example, an R^2 of **−1.8268** indicates a very bad model as the best possible score is **1**. A score of **0** can be achieved if we constantly predict the labels by using the average value of the features.

> **NOTE**
>
> We will omit the mathematical background of this score in this book.

Our model does not perform well for two reasons:

- If we check our previous Fibonacci sequence, 11 training data points and 2 testing data points are simply not enough to perform a proper predictive analysis.

- Even if we ignore the number of points, the Fibonacci sequence does not describe a linear relationship between x and y. Approximating a nonlinear function with a line is only useful if we are looking at two very close data points.

LINEAR REGRESSION USING NUMPY ARRAYS

One reason why NumPy arrays are handier than Python lists is that they can be treated as vectors. There are a few operations defined on vectors that can simplify our calculations. We can perform operations on vectors of similar lengths.

Let's take, for example, two vectors, V_1 and V_2, with three coordinates each:

V_1 = (a, b, c) with a=1, b=2, and c=3

V_2 = (d, e, f) with d=2, e=0, and f=2

The addition of these two vectors will be this:

$V_1 + V_2$ = (a+d, b+e, c+f) = (1+2, 2+0, 3+2) = (3,2,5)

The product of these two vectors will be this:

$V_1 + V_2$ = (a*d, b*e, c*f) = (1*2, 2*0, 3*2) = (2,0,6)

You can think of each vector as our datasets with, for example, the first vector as our **features set** and the second vector as our **labels set**. With Python being able to do vector calculations, this will greatly simplify the calculations required for our linear regression models.

Now, let's build a linear regression using NumPy in the following example.

Suppose we have two sets of data with 13 data points each; we want to build a linear regression that best fits all the data points for each set.

Our first set is defined as follows:

```
[2, 8, 8, 18, 25, 21, 32, 44, 32, 48, 61, 45, 62]
```

If we plot this dataset with the values (**2,8,8,18,25,21,32,44, 32,48,61,45,62**) as the *y*-axis, and the index of each value (**1,2,3,4,5,6,7,8,9,10,11,12,13**) as the *x*-axis, we will get the following plot:

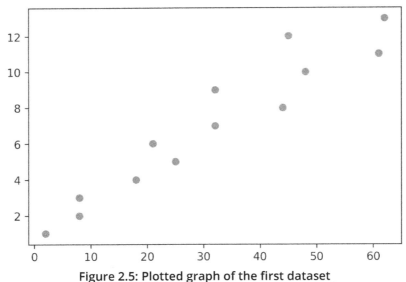

Figure 2.5: Plotted graph of the first dataset

We can see that this dataset's distribution seems linear in nature, and if we wanted to draw a line that was as close as possible to each dot, it wouldn't be too hard. A simple linear regression appears appropriate in this case.

Our second set is the first 13 values scaled in the Fibonacci sequence that we saw earlier in the *Feature Scaling* section:

```
[-0.6925069, -0.66856384, -0.66856384, -0.64462079, -0.62067773,
-0.57279161, -0.50096244, -0.38124715, -0.18970269, 0.12155706,
0.62436127, 1.43842524, 2.75529341]
```

If we plot this dataset with the values as the *y*-axis and the index of each value as the *x*-axis, we will get the following plot:

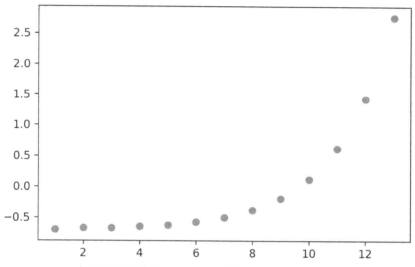

Figure 2.6: Plotted graph of the second dataset

We can see that this dataset's distribution doesn't appear to be linear, and if we wanted to draw a line that was as close as possible to each dot, our line would miss quite a lot of dots. A simple linear regression will probably struggle in this situation.

We know that the equation of a straight line is $y = a * x + b$.

In this equation, a is the slope, and b is the y intercept. To find the line of best fit, we must find the coefficients of a and b.

In order to do this, we will use the least-squares method, which can be achieved by completing the following steps:

1. For each data point, calculate x^2 and xy.

    Sum all of x, y, x^2, and $x * y$, which gives us

    $\sum X, \sum y, \sum X^2$ and $\sum Xy$ ($\sum X$ means "sum up")

2. Calculate the slope, a, as $a = \dfrac{N \sum xy - \sum x \sum y}{N \sum X^2 - (\sum X)^2}$ with N as the total number of data points.

3. Calculate the y intercept, b, as $b = \dfrac{\sum y - a \sum x}{N}$.

Now, let's apply these steps using NumPy as an example for the first dataset in the following code.

Let's take a look at the first step:

```
import numpy as np
x = np.array(range(1, 14))
y = np.array([2, 8, 8, 18, 25, 21, 32, 44, 32, 48, 61, 45, 62])
x_2 = x**2
xy = x*y
```

For **x_2**, the output will be this:

```
array([  1,   4,   9,  16,  25,  36,  49,  64,  81,
       100, 121, 144, 169],  dtype=int32)
```

For **xy**, the output will be this:

```
array([2, 16, 24, 72, 125, 126, 224,
       352, 288, 480, 671, 540, 806])
```

Now, let's move on to the next step:

```
sum_x = sum(x)
sum_y = sum(y)
sum_x_2 = sum(x_2)
sum_xy = sum(xy)
```

For **sum_x**, the output will be this:

```
91
```

For **sum_y**, the output will be this:

```
406
```

For **sum_x_2**, the output will be this:

```
819
```

For **sum_xy**, the output will be this:

```
3726
```

Now, let's move on to the next step:

```
N = len(x)
a = (N*sum_xy - (sum_x*sum_y))/(N*sum_x_2-(sum_x)**2)
```

For **N**, the output will be this:

```
13
```

For **a**, the output will be this:

```
4.857142857142857
```

Now, let's move on to the final step:

```
b = (sum_y - a*sum_x)/N
```

For **b**, the output will be this:

```
-2.7692307692307647
```

Once we plot the line $y = a * x + b$ with the preceding coefficients, we get the following graph:

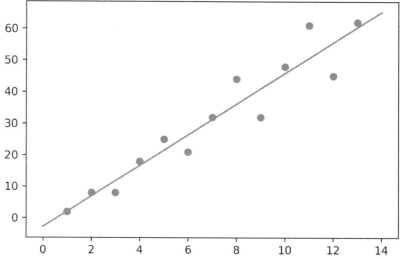

Figure 2.7: Plotted graph of the linear regression for the first dataset

As you can see, our linear regression model works quite well on this dataset, which has a linear distribution.

> **NOTE**
>
> You can find a linear regression calculator at http://www.endmemo.com/statistics/lr.php. You can also check the calculator to get an idea of what lines of best fit look like on a given dataset.

We will now repeat the exact same steps for the second dataset:

```
import numpy as np
x = np.array(range(1, 14))
y = np.array([-0.6925069, -0.66856384, -0.66856384, \
              -0.64462079, -0.62067773, -0.57279161, \
              -0.50096244, -0.38124715, -0.18970269, \
              0.12155706, 0.62436127, 1.43842524, 2.75529341])
x_2 = x**2
xy = x*y
sum_x = sum(x)
sum_y = sum(y)
sum_x_2 = sum(x_2)
sum_xy = sum(xy)
N = len(x)
a = (N*sum_xy - (sum_x*sum_y))/(N*sum_x_2-(sum_x)**2)
b = (sum_y - a*sum_x)/N
```

For **a**, the output will be this:

```
0.21838173510989017
```

For **b**, the output will be this:

```
-1.528672146538462
```

Once we plot the line $y = a * x + b$ with the preceding coefficients, we get the following graph:

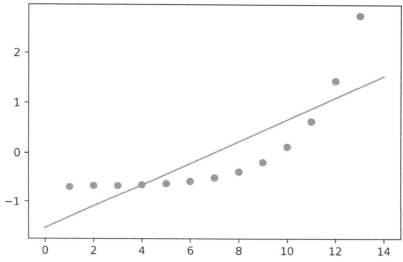

Figure 2.8: Plotted graph of the linear regression for the second dataset

Clearly, with a nonlinear distribution, our linear regression model struggles to fit the data.

NOTE

We don't have to use this method to perform linear regression. Many libraries, including scikit-learn, will help us to automate this process. Once we perform linear regression with multiple variables, we are better off using a library to perform the regression for us.

FITTING A MODEL USING NUMPY POLYFIT

NumPy Polyfit can also be used to create a line of best fit for linear regression with one variable.

Recall the calculation for the line of best fit:

```
import numpy as np
x = np.array(range(1, 14))
y = np.array([2, 8, 8, 18, 25, 21, 32, 44, 32, 48, 61, 45, 62])
x_2 = x**2
xy = x*y
sum_x = sum(x)
sum_y = sum(y)
sum_x_2 = sum(x_2)
sum_xy = sum(xy)
N = len(x)
a = (N*sum_xy - (sum_x*sum_y))/(N*sum_x_2-(sum_x)**2)
b = (sum_y - a*sum_x)/N
```

The equation for finding the coefficients a and b is quite long. Fortunately, **numpy. polyfit** in Python performs these calculations to find the coefficients of the line of best fit. The **polyfit** function accepts three arguments: the array of **x** values, the array of **y** values, and the degree of polynomial to look for. As we are looking for a straight line, the highest power of **x** is **1** in the polynomial:

```
import numpy as np
x = np.array(range(1, 14))
y = np.array([2, 8, 8, 18, 25, 21, 32, 44, 32, 48, 61, 45, 62])
[a,b] = np.polyfit(x, y, 1)
```

For **[a,b]**, the output will be this:

```
[4.857142857142858, -2.769230769230769]
```

PLOTTING THE RESULTS IN PYTHON

Suppose you have a set of data points and a regression line; our task is to plot the points and the line together so that we can see the results with our eyes.

We will use the **matplotlib.pyplot** library for this. This library has two important functions:

- **scatter**: This displays scattered points on the plane, defined by a list of *x* coordinates and a list of *y* coordinates.

- **plot**: Along with two arguments, this function plots a segment defined by two points or a sequence of segments defined by multiple points. A plot is like a scatter, except that instead of displaying the points, they are connected by lines.

A plot with three arguments plots a segment and/or two points formatted according to the third argument.

A segment is defined by two points. As *x* ranges between 1 and 13 (remember the dataset contains 13 data points), it makes sense to display a segment between 0 and 15. We must substitute the value of *x* in the equation $a * x + b$ to get the corresponding *y* values:

```
import numpy as np
import matplotlib.pyplot as plot
x = np.array(range(1, 14))
y = np.array([2, 8, 8, 18, 25, 21, 32, 44, 32, 48, 61, 45, 62])
x_2 = x**2
xy = x*y
sum_x = sum(x)
sum_y = sum(y)
sum_x_2 = sum(x_2)
sum_xy = sum(xy)
N = len(x)
a = (N*sum_xy - (sum_x*sum_y))/(N*sum_x_2-(sum_x)**2)
b = (sum_y - a*sum_x)/N

# Plotting the points
plot.scatter(x, y)
# Plotting the line
plot.plot([0, 15], [b, 15*a+b])
plot.show()
```

The output is as follows:

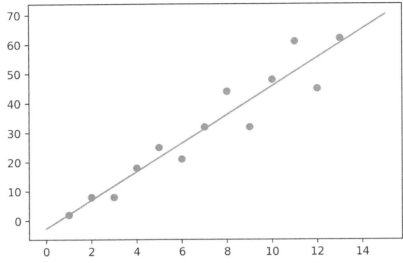

Figure 2.9: Plotted graph of the linear regression for the first dataset using matplotlib

The regression line and the scattered data points are displayed as expected.

However, the plot has an advanced signature. You can use **plot** to draw scattered dots, lines, and any curves on this figure. These variables are interpreted in groups of three:

- **x** values

- **y** values

- Formatting options in the form of a string

Let's create a function for deriving an array of approximated **y** values from an array of approximated **x** values:

```
def fitY( arr ):
    return [4.857142857142859 * x - 2.7692307692307843 for x in arr]
```

We will use the **fit** function to plot the values:

```
plot.plot(x, y, 'go',x, fitY(x), 'r--o')
```

Every third argument handles formatting. The letter **g** stands for green, while the letter **r** stands for red. You could have used **b** for blue and **y** for yellow, among other examples. In the absence of a color, each triple value will be displayed using a different color. The **o** character symbolizes that we want to display a dot where each data point lies. Therefore, **go** has nothing to do with movement – it requests the plotter to plot green dots. The **–** characters are responsible for displaying a dashed line. If you just use -1, a straight line appears instead of the dashed line.

The output is as follows:

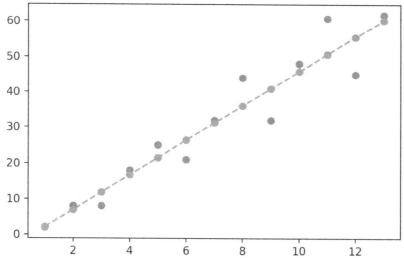

Figure 2.10: Graph for the plot function using the fit function

The Python plotter library offers a simple solution for most of your graphing problems. You can draw as many lines, dots, and curves as you want on this graph.

When displaying curves, the plotter connects the dots with segments. Also, bear in mind that even a complex sequence of curves is an approximation that connects the dots. For instance, if you execute the code from https://gist.github. com/traeblain/1487795, you will recognize the segments of the **batman** function as connected lines:

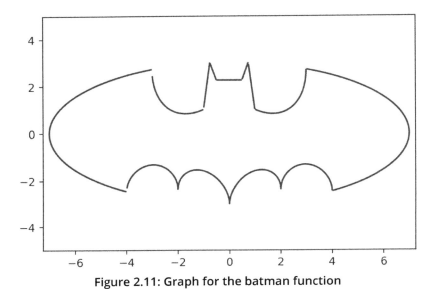

Figure 2.11: Graph for the batman function

There is a large variety of ways to plot curves. We have seen that the **polyfit** method of the NumPy library returns an array of coefficients to describe a linear equation:

```
import numpy as np
x = np.array(range(1, 14))
y = np.array([2, 8, 8, 18, 25, 21, 32, 44, 32, 48, 61, 45, 62])
np.polyfit(x, y, 1)
```

Here the output is as follows:

```
[4.857142857142857, -2.769230769230768]
```

This array describes the equation *4.85714286 * x - 2.76923077*.

Suppose we now want to plot a curve, $y = -x^2 + 3x - 2$. This quadratic equation is described by the coefficient array **[-1, 3, -2]** as $y = (-1)*x^2 + (3)*x - 2$. We could write our own function to calculate the **y** values belonging to **x** values. However, the NumPy library already has a feature that can do this work for us – **np.poly1d**:

```
import numpy as np
x = np.array(range( -10, 10, 1 ))
f = np.poly1d([-1,3,-2])
```

The **f** function that's created by the **poly1d** call not only works with single values but also with lists or NumPy arrays:

```
f(5)
```

The expected output is this:

```
-12
```

Similarly, for **f(x)**:

```
f(x)
```

The output will be:

```
array ([-132. -110, -90, -72, -56, -42, -30, -20, -12, -6, -2,
        0, 0, -2, -6, -12, -20, -30, -42, -56])
```

We can now use these values to plot a nonlinear curve:

```
import matplotlib.pyplot as plot
plot.plot(x, f(x))
```

The output is as follows:

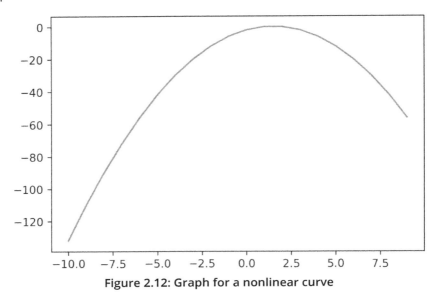

Figure 2.12: Graph for a nonlinear curve

As you can see, we can use the **pyplot** library to easily create the plot of a nonlinear curve.

PREDICTING VALUES WITH LINEAR REGRESSION

Suppose we are interested in the **y** value belonging to the **x** coordinate **20**. Based on the linear regression model, all we need to do is substitute the value of **20** in the place of **x** on the previously used code:

```
x = np.array(range(1, 14))
y = np.array([2, 8, 8, 18, 25, 21, 32, 44, 32, 48, 61, 45, 62])
# Plotting the points
plot.scatter(x, y)
# Plotting the prediction belonging to x = 20
plot.scatter(20, a * 20 + b, color='red')
# Plotting the line
plot.plot([0, 25], [b, 25*a+b])
```

The output is as follows:

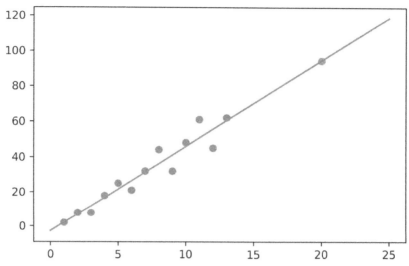

Figure 2.13: Graph showing the predicted value using linear regression

Here, we denoted the predicted value with red. This red point is on the best line of fit.

Let's look at next exercise where we will be predicting populations based on linear regression.

EXERCISE 2.01: PREDICTING THE STUDENT CAPACITY OF AN ELEMENTARY SCHOOL

In this exercise, you will be trying to forecast the need for elementary school capacity. Your task is to figure out 2025 and 2030 predictions for the number of children starting elementary school.

> **NOTE**
>
> The data is contained inside the **population.csv** file, which you can find on our GitHub repository: https://packt.live/2YYlPoj.

The following steps will help you to complete this exercise:

1. Open a new Jupyter Notebook file.

2. Import **pandas** and **numpy**:

```
import pandas as pd
import numpy as np
import matplotlib.pyplot as plot
```

3. Next, load the CSV file as a DataFrame on the Notebook and read the CSV file:

> **NOTE**
>
> Watch out for the slashes in the string below. Remember that the backslashes (\ ) are used to split the code across multiple lines, while the forward slashes (/) are part of the URL.

```
file_url = 'https://raw.githubusercontent.com/'\
           'PacktWorkshops/The-Applied-Artificial-'\
           'Intelligence-Workshop/master/Datasets/'\
           'population.csv'
df = pd.read_csv(file_url)
df
```

The expected output is this:

	year	population
0	2001	147026
1	2002	144272
2	2003	140020
3	2004	143801
4	2005	146233
5	2006	144539
6	2007	141273
7	2008	135389
8	2009	142500
9	2010	139452
10	2011	139722
11	2012	135300
12	2013	137289
13	2014	136511
14	2015	132884
15	2016	125683
16	2017	127255
17	2018	124275

Figure 2.14: Reading the CSV file

4. Now, convert the DataFrame into two NumPy arrays. For simplicity, we can indicate that the **year** feature, which is from **2001** to **2018**, is the same as **1** to **18**:

```
x = np.array(range(1, 19))
y = np.array(df['population'])
```

The **x** output will be:

```
array([1, 2, 3, 4, 5, 6, 7, 8, 9, 10, 11, 12, 13, 14, 15, 16, 17, 18])
```

The **y** output will be:

```
array([147026, 144272, 140020, 143801, 146233,
       144539, 141273, 135389, 142500, 139452,
       139722, 135300, 137289, 136511, 132884,
       125683, 127255, 124275], dtype=int64)
```

5. Now, with the two NumPy arrays, use the **polyfit** method (with a degree of **1** as we only have one feature) to determine the coefficients of the regression line:

```
[a, b] = np.polyfit(x, y, 1)
```

The output for **[a, b]** will be:

```
[-1142.0557275541803, 148817.5294117647]
```

6. Now, plot the results using **matplotlib.pyplot** and predict the future until **2030**:

```
plot.scatter( x, y )
plot.plot( [0, 30], [b, 30*a+b] )
plot.show()
```

The expected output is this:

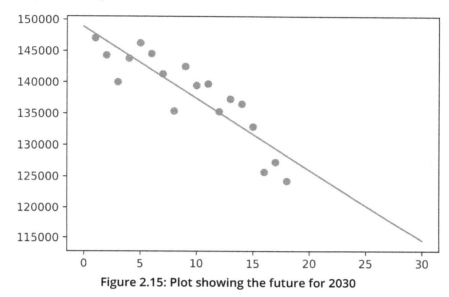

Figure 2.15: Plot showing the future for 2030

As you can see, the data appears linear and our model seems to be a good fit.

7. Finally, predict the population for **2025** and **2030**:

```
population_2025 = 25*a+b
population_2030 = 30*a+b
```

The output for **population_2025** will be:

```
120266.1362229102
```

The output for **population_2030** will be:

```
114555.85758513928
```

> **NOTE**
>
> To access the source code for this specific section, please refer to https://packt.live/31dvuKt.
>
> You can also run this example online at https://packt.live/317qelc. You must execute the entire Notebook in order to get the desired result.

By completing this exercise, we can now conclude that the population of children starting elementary school is going to decrease in the future and that there is no need to increase the elementary school capacity if we are currently meeting the needs.

LINEAR REGRESSION WITH MULTIPLE VARIABLES

In the previous section, we dealt with linear regression with one variable. Now we will learn an extended version of linear regression, where we will use multiple input variables to predict the output.

MULTIPLE LINEAR REGRESSION

If you recall the formula for the line of best fit in linear regression, it was defined as $y = a * x + b$, where a is the slope of the line, b is the y intercept of the line, x is the feature value, and y is the calculated label value.

In multiple regression, we have multiple features and one label. If we have three features, x_1, x_2, and x_3, our model changes to $y = a_1 * x_1 + a_2 * x_2 + a_3 * x_3 + b$.

In NumPy array format, we can write this equation as follows:

```
y = np.dot(np.array([a1, a2, a3]), np.array([x1, x2, x3])) + b
```

For convenience, it makes sense to define the whole equation in a vector multiplication format. The coefficient of b is going to be **1**:

```
y = np.dot(np.array([b, a1, a2, a3]) * np.array([1, x1, x2, x3]))
```

Multiple linear regression is a simple scalar product of two vectors, where the coefficients b, a_1, a_2, and a_3 determine the best fit equation in a four-dimensional space.

To understand the formula of multiple linear regression, you will need the scalar product of two vectors. As the other name for a scalar product is a dot product, the NumPy function performing this operation is called **dot**:

```
import numpy as np
v1 = [1, 2, 3]
v2 = [4, 5, 6]
np.dot(v1, v2)
```

The output will be **32** as **np.dot(v1, v2) = 1 * 4 + 2 * 5 + 3 * 6 = 32**.

We simply sum the product of each respective coordinate.

We can determine these coefficients by minimizing the error between the data points and the nearest points described by the equation. For simplicity, we will omit the mathematical solution of the best-fit equation and use scikit-learn instead.

> **NOTE**
>
> In n-dimensional spaces, where n is greater than 3, the number of dimensions determines the different variables that are in our model. In the preceding example, we have three features (x_1, x_2, and x_3) and one label, y. This yields four dimensions. If you want to imagine a four-dimensional space, you can imagine a three-dimensional space with a fourth dimension of time. A five-dimensional space can be imagined as a four-dimensional space, where each point in time has a temperature. Dimensions are just features (and labels); they do not necessarily correlate with our concept of three-dimensional space.

THE PROCESS OF LINEAR REGRESSION

We will follow the following simple steps to solve linear regression problems:

1. Load data from the data sources.

2. Prepare data for prediction. Data is prepared in this (**normalize**, **format**, and **filter**) format.

3. Compute the parameters of the regression line. Regardless of whether we use linear regression with one variable or with multiple variables, we will follow these steps.

IMPORTING DATA FROM DATA SOURCES

There are multiple libraries that can provide us with access to data sources. As we will be working with stock data, let's cover two examples that are geared toward retrieving financial data: Quandl and Yahoo Finance. Take a look at these important points before moving ahead:

*   Scikit-learn comes with a few datasets that can be used for practicing your skills.

*   https://www.quandl.com provides you with free and paid financial datasets.

*   https://pandas.pydata.org/ helps you load any CSV, Excel, JSON, or SQL data.

*   Yahoo Finance provides you with financial datasets.

LOADING STOCK PRICES WITH YAHOO FINANCE

The process of loading stock data with Yahoo Finance is straightforward. All you need to do is install the **yfinance** package using the following command in Jupyter Notebook:

```
!pip install yfinance
```

We will download a dataset that has an open price, high price, low price, close price, adjusted close price, and volume values of the S&P 500 index starting from 2015 to January 1, 2020. The S&P 500 index is the stock market index that measures the stock performance of 500 large companies listed in the United States:

```
import yfinance as yahoo
spx_data_frame = yahoo.download("^GSPC", "2015-01-01", "2020-01-01")
```

> **NOTE**
>
> The dataset file can also be found in our GitHub repository: https://packt.live/3fRI5Hk.
>
> The original dataset can be found here: https://github.com/ranaroussi/yfinance.

That's all you need to do. The DataFrame containing the S&P 500 index is ready.

You can plot the index closing prices using the **plot** method:

```
spx_data_frame.Close.plot()
```

The output is as follows:

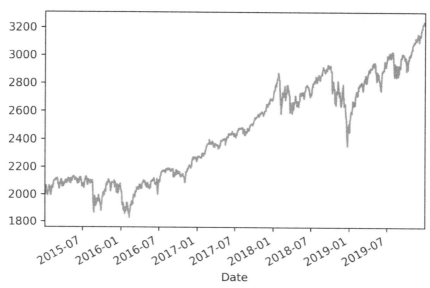

Figure 2.16: Graph showing the S&P 500 index closing price since 2015

The data does not appear to be linear; a polynomial regression might be a better model for this dataset.

It is also possible to save data to a CSV file using the following code:

```
spx_data_frame.to_csv("yahoo_spx.csv")
```

> **NOTE**
>
> https://www.quandl.com is a reliable source of financial and economic datasets that we will be using in this chapter.

EXERCISE 2.02: USING QUANDL TO LOAD STOCK PRICES

The goal of this exercise is to download data from the Quandl package and load it into a DataFrame like we previously did with Yahoo Finance.

The following steps will help you to complete the exercise:

1. Open a new Jupyter Notebook file.

2. Install **Quandl** using the following command:

```
!pip install quandl
```

3. Download the data into a DataFrame using Quandl for the S&P 500. Its ticker is **"YALE/SPCOMP"**:

```
import quandl
data_frame = quandl.get("YALE/SPCOMP")
```

4. Use the DataFrame **head()** method to inspect the first five rows of data in your DataFrame:

```
data_frame.head()
```

The output is as follows:

Year	S&P Composite	Dividend	Earnings	CPI	Long Interest Rate	Real Price	Real Dividend	Real Earnings	Cyclically Adjusted PE Ratio
1871-01-31	4.44	0.26	0.4	12.464061	5.320000	91.599130	5.363913	8.252174	NaN
1871-02-28	4.50	0.26	0.4	12.844641	5.323333	90.086245	5.204983	8.007666	NaN
1871-03-31	4.61	0.26	0.4	13.034972	5.326667	90.940801	5.128982	7.890742	NaN
1871-04-30	4.74	0.26	0.4	12.559226	5.330000	97.047287	5.323269	8.189645	NaN
1871-05-31	4.86	0.26	0.4	12.273812	5.333333	101.818048	5.447056	8.380086	NaN

Figure 2.17: Dataset displayed as the output

> **NOTE**
>
> To access the source code for this specific section, please refer to https://packt.live/3dwDUz6.
>
> You can also run this example online at https://packt.live/31812B6. You must execute the entire Notebook in order to get the desired result.

By completing this exercise, we have learned how to download an external dataset in **CSV** format and import it as a DataFrame. We also learned about the `.head()` method, which provides a quick view of the first five rows of your DataFrame.

In the next section, we will be moving on to prepare the dataset to perform multiple linear regression.

PREPARING DATA FOR PREDICTION

Before we perform multiple linear regression on our dataset, we must choose the relevant features and the data range on which we will perform the regression.

Preparing the data for prediction is the second step in the regression process. This step also has several sub-steps. We will go through these sub-steps in the following exercise.

EXERCISE 2.03: PREPARING THE QUANDL DATA FOR PREDICTION

The goal of this exercise is to download an external dataset from the Quandl library and then prepare it so that it is ready for use in our linear regression models.

The following steps will help you to complete this exercise:

1. Open a new Jupyter Notebook file.

> **NOTE**
>
> If the Qaundl library is not installed on your system, remember to run the command `!pip install quandl`.

2. Next, download the data into a DataFrame using Quandl for the S&P 500 between 1950 and 2019. Its ticker is **"YALE/SPCOMP"**:

```
import quandl
import numpy as np
from sklearn import preprocessing
from sklearn import model_selection
data_frame = quandl.get("YALE/SPCOMP", \
                    start_date="1950-01-01", \
                    end_date="2019-12-31")
```

3. Use the **head()** method to visualize the columns inside the **data_frame.head()** DataFrame:

```
data_frame.head()
```

The output is as follows:

Year	S&P Composite	Dividend	Earnings	CPI	Long Interest Rate	Real Price	Real Dividend	Real Earnings	Cyclically Adjusted PE Ratio
1950-01-31	16.88	1.15	2.33667	23.5	2.320000	184.702397	12.583398	25.568042	10.745733
1950-02-28	17.21	1.16	2.35333	23.5	2.340833	188.313285	12.692819	25.750337	10.911564
1950-03-31	17.35	1.17	2.37000	23.6	2.361667	189.040748	12.747993	25.822857	10.910947
1950-04-30	17.84	1.18	2.42667	23.6	2.382500	194.379651	12.856950	26.440318	11.178022
1950-05-31	18.44	1.19	2.48333	23.7	2.403333	200.069332	12.911199	26.943502	11.461543

Figure 2.18: Dataset displayed as the output

A few features seem to highly correlate with each other. For instance, the **Real Dividend** column grows proportionally with **Real Price**. The ratio between them is not always similar, but they do correlate.

As regression is not about detecting the correlation between features, we would rather get rid of the features that we know are correlated and perform regression on the features that are non-correlated. In this case, we will keep the **Long Interest Rate**, **Real Price**, and **Real Dividend** columns.

4. Keep only the relevant columns in the **Long Interest Rate**, **Real Price**, and **Real Dividend** DataFrames:

```
data_frame = data_frame[['Long Interest Rate', \
                          'Real Price', 'Real Dividend']]
data_frame
```

The output is as follows:

Year	Long Interest Rate	Real Price	Real Dividend
1950-01-31	2.320000	184.702397	12.583398
1950-02-28	2.340833	188.313285	12.692819
1950-03-31	2.361667	189.040748	12.747993
1950-04-30	2.382500	194.379651	12.856950
1950-05-31	2.403333	200.069332	12.911199
...	...	...	...
2019-08-31	1.630000	2904.059842	56.967809
2019-09-30	1.700000	2986.569552	57.304685
2019-10-31	1.710000	2975.284860	NaN
2019-11-30	1.810000	3104.071562	NaN
2019-12-31	1.910000	3223.380000	NaN

840 rows × 3 columns

Figure 2.19: Dataset showing only the relevant columns

You can see that the DataFrame contains a few missing values **NaN**. As regression doesn't work with missing values, we need to either replace them or delete them. In the real world, we will usually choose to replace them. In this case, we will replace the missing values by the preceding values using a method called **forward filling**.

5. We can replace the missing values with a forward filling as shown in the following code snippet:

```
data_frame.fillna(method='ffill', inplace=True)
data_frame
```

The output is as follows:

Year	Long Interest Rate	Real Price	Real Dividend
1950-01-31	2.320000	184.702397	12.583398
1950-02-28	2.340833	188.313285	12.692819
1950-03-31	2.361667	189.040748	12.747993
1950-04-30	2.382500	194.379651	12.856950
1950-05-31	2.403333	200.069332	12.911199
...	...	...	...
2019-08-31	1.630000	2904.059842	56.967809
2019-09-30	1.700000	2986.569552	57.304685
2019-10-31	1.710000	2975.284860	57.304685
2019-11-30	1.810000	3104.071562	57.304685
2019-12-31	1.910000	3223.380000	57.304685

840 rows × 3 columns

Figure 2.20: Missing values have been replaced

Now that we have cleaned the missing data, we need to create our label. We want to predict the **Real Price** column 3 months in advance using the current **Real Price**, **Long Interest Rate**, and **Real Dividend** columns. In order to create our label, we need to shift the **Real Price** values up by three units and call it **Real Price Label**.

6. Create the **Real Price Label** label by shifting **Real Price** by 3 months as shown in the following code:

```
data_frame['Real Price Label'] = data_frame['Real Price'].shift(-3)
data_frame
```

The output is as follows:

Year	Long Interest Rate	Real Price	Real Dividend	Real Price Label
1950-01-31	2.320000	184.702397	12.583398	194.379651
1950-02-28	2.340833	188.313285	12.692819	200.069332
1950-03-31	2.361667	189.040748	12.747993	202.469952
1950-04-30	2.382500	194.379651	12.856950	185.438831
1950-05-31	2.403333	200.069332	12.911199	195.023530
...	...	...	...	...
2019-08-31	1.630000	2904.059842	56.967809	3104.071562
2019-09-30	1.700000	2986.569552	57.304685	3223.380000
2019-10-31	1.710000	2975.284860	57.304685	NaN
2019-11-30	1.810000	3104.071562	57.304685	NaN
2019-12-31	1.910000	3223.380000	57.304685	NaN

840 rows × 4 columns

Figure 2.21: New labels have been created

The side effect of shifting these values is that missing values will appear in the last three rows for **Real Price Label**, so we need to remove the last three rows of data. However, before that, we need to convert the features into a NumPy array and scale it. We can use the **drop** method of the DataFrame to remove the label column and the preprocessing function from **sklearn** to scale the features.

7. Create a NumPy array for the features and scale it in the following code:

```
features = np.array(data_frame.drop('Real Price Label', 1))
scaled_features = preprocessing.scale(features)
scaled_features
```

The output is as follows:

```
array([[-1.14839975, -1.13009904, -1.19222544],
       [-1.14114523, -1.12483455, -1.18037146],
       [-1.13389072, -1.12377394, -1.17439424],
       ...,
       [-1.360812  ,  2.9384288 ,  3.65260385],
       [-1.32599032,  3.12619329,  3.65260385],
       [-1.29116864,  3.30013894,  3.65260385]])
```

The **1** in the second argument specifies that we are dropping columns. As the original DataFrame was not modified, the label can be directly extracted from it. Now that the features are scaled, we need to remove the last three values of the features as they are the features of the missing values in the label column. We will save them for later in the prediction part.

8. Remove the last three values of the **features** array and save them into another array using the following code:

```
scaled_features_latest_3 = scaled_features[-3:]
scaled_features = scaled_features[:-3]
scaled_features
```

The output for **scaled_features** is as follows:

```
array([[-1.14839975, -1.13009904, -1.19222544],
       [-1.14114523, -1.12483455, -1.18037146],
       [-1.13389072, -1.12377394, -1.17439424],
       ...,
       [-1.38866935,  2.97846643,  3.57443947],
       [-1.38866935,  2.83458633,  3.6161088 ],
       [-1.36429417,  2.95488131,  3.65260385]])
```

The **scaled_features** variable doesn't contain the three data points anymore as they are now in **scaled_features_latest_3**. Now we can remove the last three rows with missing data from the DataFrame, then convert the label into a NumPy array using **sklearn**.

9. Remove the rows with missing data in the following code:

```
data_frame.dropna(inplace=True)
data_frame
```

The output for **data_frame** is as follows:

Year	Long Interest Rate	Real Price	Real Dividend	Real Price Label
1950-01-31	2.320000	184.702397	12.583398	194.379651
1950-02-28	2.340833	188.313285	12.692819	200.069332
1950-03-31	2.361667	189.040748	12.747993	202.469952
1950-04-30	2.382500	194.379651	12.856950	185.438831
1950-05-31	2.403333	200.069332	12.911199	195.023530
...	...	...	...	...
2019-05-31	2.400000	2866.381124	55.925898	2904.059842
2019-06-30	2.060000	2901.408290	56.295329	2986.569552
2019-07-31	1.630000	3002.746469	56.583171	2975.284860
2019-08-31	1.630000	2904.059842	56.967809	3104.071562
2019-09-30	1.700000	2986.569552	57.304685	3223.380000

837 rows × 4 columns

Figure 2.22: Dataset updated with the removal of missing values

As you can see, the last three rows were also removed from the DataFrame.

10. Now let's see if we have accurately created our label. Go ahead and run the following code:

```
label = np.array(data_frame['Real Price Label'])
label
```

The output for the **label** is as follows:

```
array([ 194.37965085,  200.06933165,  202.4699521 ,  185.43883071,
        195.02352963,  201.07426721,  207.69723293,  206.43993401,
        203.13981    ,  214.72118858,  220.11898833,  215.57816163,
        218.46848372,  217.72425753,  213.95156178,  217.72425753,
        227.25527838,  231.32657931,  229.26591756,  221.19798068,
        227.15562226,  234.72424189,  232.20727186,  232.79390076,
        231.23029773,  231.13289659,  236.56788    ,  241.53730787,
        242.50037528,  238.64810562,  233.64015506,  241.05577416,
        250.78275506,  253.07891053,  250.92885057,  251.24220338,
        238.86859737,  239.22594607,  229.7939944 ,  233.0562056 ,
        233.14573271,  222.43957361,  228.28229    ,  234.19723048,
        237.35172379,  243.37393829,  248.72701784,  253.98450669,
        265.10263321,  274.63209926,  276.8306855 ,  288.01479814,
        293.75024052,  301.75453545,  308.75869478,  320.84806567,
        336.78467528,  342.852      ,  354.31250225,  351.51960674,
        363.65425618,  362.11334831,  383.10821798,  409.59939963,
        407.10476754,  423.84919182,  402.53246431,  429.68022491,
        435.31329963,  423.60771828,  426.29424515,  455.6541459 ,
        459.31334387,  443.23144667,  437.41990919,  457.78249708,
```

Figure 2.23: Output showing the expected labels

Our variable contains all the labels and is exactly the same as the **Real Price Label** column in the DataFrame.

Our next task is to separate the training and testing data from each other. As we saw in the *Splitting Data into Training and Testing* section, we will use 90% of the data as the training data and the remaining 10% as the test data.

11. Split the **features** data into training and test sets using **sklearn** with the following code:

```
from sklearn import model_selection
(features_train, features_test, \
label_train, label_test) = model_selection\
                    .train_test_split(scaled_features, \
                                      label, test_size=0.1, \
                                      random_state=8)
```

The **train_test_split** function shuffles the lines of our data, keeps the correspondence, and puts approximately 10% of all data in the test variables, keeping 90% for the training variables. We also use **random_state=8** in order to reproduce the results. Our data is now ready to be used for the multiple linear regression model.

> **NOTE**
>
> To access the source code for this specific section, please refer to https://packt.live/2zZssOG.
>
> You can also run this example online at https://packt.live/2zW8WCH. You must execute the entire Notebook in order to get the desired result.

By completing this exercise, we have learned all the required steps for data preparation before performing a regression.

PERFORMING AND VALIDATING LINEAR REGRESSION

Now that our data has been prepared, we can perform our linear regression. After that, we will measure our model performance and see how well it performs.

We can now create the linear regression model based on the training data:

```
from sklearn import linear_model
model = linear_model.LinearRegression()
model.fit(features_train, label_train)
```

Once the model is ready, we can use it to predict the labels belonging to the test feature values and use the **score** method from the model to see how accurate it is:

```
label_predicted = model.predict(features_test)
model.score(features_test, label_test)
```

The output is as follows:

```
0.9847223874806746
```

With a score or R^2 of **0.985**, we can conclude that the model is very accurate. This is not a surprise since the financial market grows at around 6-7% a year. This is linear growth, and the model essentially predicts that the markets will continue growing at a linear rate. Concluding that markets tend to increase in the long run is not rocket science.

PREDICTING THE FUTURE

Now that our model has been trained, we can use it to predict future values. We will use the **scaled_features_latest_3** variable that we created by taking the last three values of the features NumPy array and using it to predict the index price of the next three months in the following code:

```
label_predicted = model.predict(scaled_features_latest_3)
```

The output is as follows:

```
array ([3046.2347327, 3171.47495182, 3287.48258298])
```

By looking at the output, you might think it seems easy to forecast the value of the S&P 500 and use it to earn money by investing in it. Unfortunately, in practice, using this model for making money by betting on the forecast is by no means better than gambling in a casino. This is just an example to illustrate prediction; it is not enough to be used for short-term or long-term speculation on market prices. In addition to this, stock prices are sensitive to many external factors, such as economic recession and government policy. This means that past patterns do not necessarily reflect any patterns in the future.

POLYNOMIAL AND SUPPORT VECTOR REGRESSION

When performing a polynomial regression, the relationship between x and y, or using their other names, features, and labels, is not a linear equation, but a polynomial equation. This means that instead of the $y = a * x + b$ equation, we can have multiple coefficients and multiple powers of x in the equation.

To make matters even more complicated, we can perform polynomial regression using multiple variables, where each feature may have coefficients multiplying different powers of the feature.

Our task is to find a curve that best fits our dataset. Once polynomial regression is extended to multiple variables, we will learn the SVM model to perform polynomial regression.

POLYNOMIAL REGRESSION WITH ONE VARIABLE

As a recap, we have performed two types of regression so far:

- Simple linear regression: $y = a * x + b$
- Multiple linear regression: $y = b + a_1 * x_1 + a_2 * x_2 + \ldots + a_n * x_n$

We will now learn how to do polynomial linear regression with one variable. The equation for polynomial linear regression is

$$y = b + a_1 * x + a_2 * (x)^2 + a_3 * (x)^3 + \ldots + a_n * (x)^n.$$

Polynomial linear regression has a vector of coefficients, $(b, a_1, a_2, \ldots, a_n)$, multiplying a vector of degrees of x in the polynomial, $(1, x^1, x^2, \ldots, x^n)$.

At times, polynomial regression works better than linear regression. If the relationship between labels and features can be described using a linear equation, then using a linear equation makes perfect sense. If we have a nonlinear growth, polynomial regression tends to approximate the relationship between features and labels better.

The simplest implementation of linear regression with one variable was the `polyfit` method of the NumPy library. In the next exercise, we will perform multiple polynomial linear regression with degrees of 2 and 3.

> **NOTE**
>
> Even though our polynomial regression has an equation containing coefficients of x^n, this equation is still referred to as polynomial linear regression in literature. Regression is made linear not because we restrict the usage of higher powers of x in the equation, but because the coefficients $a_1, a_2 \ldots$ and so on are linear in the equation. This means that we use the toolset of linear algebra and work with matrices and vectors to find the missing coefficients that minimize the error of the approximation.

EXERCISE 2.04: FIRST-, SECOND-, AND THIRD-DEGREE POLYNOMIAL REGRESSION

The goal of this exercise is to perform first-, second-, and third-degree polynomial regression on the two sample datasets that we used earlier in this chapter. The first dataset has a linear distribution and the second one is the Fibonacci sequence and has a nonlinear distribution.

The following steps will help you to complete the exercise:

1. Open a new Jupyter Notebook file.

2. Import the **numpy** and **matplotlib** packages:

```
import numpy as np
from matplotlib import pyplot as plot
```

3. Define the first dataset:

```
x1 = np.array(range(1, 14))
y1 = np.array([2, 8, 8, 18, 25, 21, 32, \
                44, 32, 48, 61, 45, 62])
```

4. Define the second dataset:

```
x2 = np.array(range(1, 14))
y2 = np.array([0, 1, 1, 2, 3, 5, 8, 13, \
                21, 34, 55, 89, 144])
```

5. Perform a polynomial regression of degrees **1**, **2**, and **3** on the first dataset using the **polyfit** method from **numpy** in the following code:

```
f1 = np.poly1d(np.polyfit(x1, y1, 1))
f2 = np.poly1d(np.polyfit(x1, y1, 2))
f3 = np.poly1d(np.polyfit(x1, y1, 3))
```

The output for **f1** is as follows:

```
poly1d([ 4.85714286, -2.76923077])
```

As you can see, a polynomial regression of degree **1** has two coefficients.

The output for **f2** is as follows:

```
poly1d([-0.03196803, 5.3046953, -3.88811189])
```

As you can see, a polynomial regression of degree **2** has three coefficients.

The output for **f3** is as follows:

```
poly1d([-0.01136364, 0.20666833, -3.91833167, -1.97902098])
```

As you can see, a polynomial regression of degree **3** has four coefficients.

Now that we have calculated the three polynomial regressions, we can plot them together with the data on a graph to see how they behave.

6. Plot the three polynomial regressions and the data on a graph in the following code:

```
import matplotlib.pyplot as plot
plot.plot(x1, y1, 'ko', # black dots \
            x1, f1(x1),'k-',  # straight line \
            x1, f2(x1),'k--',  # black dashed line \
```

```
        x1,  f3(x1),'k-.'  # dot line
)
plot.show()
```

The output is as follows:

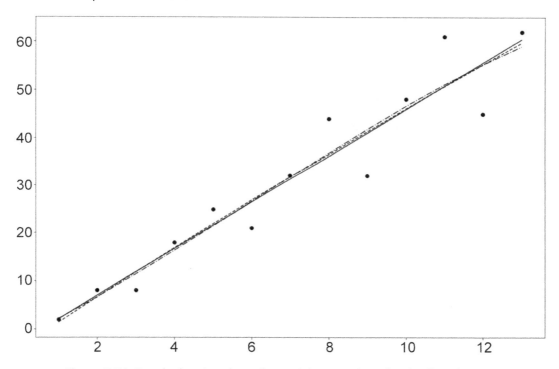

Figure 2.24: Graph showing the polynomial regressions for the first dataset

As the coefficients are enumerated from left to right in order of decreasing degree, we can see that the higher-degree coefficients stay close to negligible. In other words, the three curves are almost on top of each other, and we can only detect a divergence near the right edge. This is because we are working on a dataset that can be very well approximated with a linear model.

In fact, the first dataset was created out of a linear function. Any non-zero coefficients for x^2 and x^3 are the result of overfitting the model based on the available data. The linear model is better for predicting values outside the range of the training data than any higher-degree polynomial.

Let's contrast this behavior with the second example. We know that the Fibonacci sequence is nonlinear. So, using a linear equation to approximate it is a clear case for underfitting. Here, we expect a higher polynomial degree to perform better.

7. Perform a polynomial regression of degrees **1**, **2**, and **3** on the second dataset using the **polyfit** method from **numpy** with the following code:

```
g1 = np.poly1d(np.polyfit(x2, y2, 1))
g2 = np.poly1d(np.polyfit(x2, y2, 2))
g3 = np.poly1d(np.polyfit(x2, y2, 3))
```

The output for **g1** is as follows:

```
poly1d([ 9.12087912, -34.92307692])
```

As you can see, a polynomial regression of degree **1** has **2** coefficients.

The output for **g2** is as follows:

```
poly1d([ 1.75024975, -15.38261738, 26.33566434])
```

As you can see, a polynomial regression of degree **2** has **3** coefficients.

The output for **g3** is as follows:

```
poly1d([ 0.2465035, -3.42632368, 14.69080919, -15.07692308])
```

As you can see, a polynomial regression of degree **3** has **4** coefficients.

8. Plot the three polynomial regressions and the data on a graph in the following code:

```
plot.plot(x2, y2, 'ko', # black dots \
          x2, g1(x2),'k-',  # straight line \
          x2, g2(x2),'k--',  # black dashed line \
          x2, g3(x2),'k-.' # dot line
)
plot.show()
```

The output is as follows:

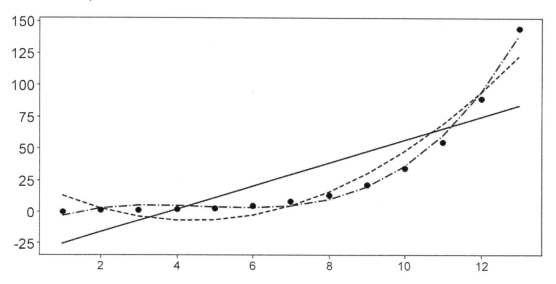

Figure 2.25: Graph showing the second dataset points and three polynomial curves

The difference is clear. The quadratic curve fits the points a lot better than the linear one. The cubic curve is even better.

> **NOTE**
>
> To access the source code for this specific section, please refer to https://packt.live/3dpCgyY.
>
> You can also run this example online at https://packt.live/2B09xDN. You must execute the entire Notebook in order to get the desired result.

If you research Binet's formula, you will find out that the Fibonacci function is an exponential function, as the nth Fibonacci number is calculated as the nth power of a constant. Therefore, the higher the polynomial degree we use, the more accurate our approximation will be.

POLYNOMIAL REGRESSION WITH MULTIPLE VARIABLES

When we have one variable of degree *n*, we have *n+1* coefficients in the equation as

$$y = b + a_1 * x + a_2 * (x)^2 + a_3 * (x)^3 + \ldots + a_n * (x)^n.$$

Once we deal with multiple features, x^1, x^2, ..., x^m, and their powers of up to the nth degree, we get an *m * (n+1)* matrix of coefficients. The math will become quite lengthy when we start exploring the details and prove how a polynomial model works. We will also lose the nice visualizations of two-dimensional curves.

Therefore, we will apply the concepts learned in the previous section on polynomial regression with one variable and omit the math. When training and testing a linear regression model, we can calculate the mean square error to see how good an approximation a model is.

In scikit-learn, the degree of the polynomials used in the approximation is a simple parameter in the model.

As polynomial regression is a form of linear regression, we can perform polynomial regression without changing the regression model. All we need to do is to transform the input and keep the linear regression model. The transformation of the input is performed by the **fit_transform** method of the **PolynomialFeatures** package.

First, we can reuse the code from *Exercise 2.03, Preparing the Quandl Data for Prediction*, up to *Step 9* and import **PolynomialFeatures** from the **preprocessing** module of **sklearn**:

```
!pip install quandl
import quandl
import numpy as np
from sklearn import preprocessing
from sklearn import model_selection
from sklearn import linear_model
from matplotlib import pyplot as plot
from sklearn.preprocessing import PolynomialFeatures

data_frame = quandl.get("YALE/SPCOMP", \
                start_date="1950-01-01", \
                end_date="2019-12-31")
data_frame = data_frame[['Long Interest Rate', \
                'Real Price', 'Real Dividend']]
```

```
data_frame.fillna(method='fill', inplace=True)
data_frame['Real Price Label'] = data_frame['Real Price'].shift(-3)
features = np.array(data_frame.drop('Real Price Label', 1))
scaled_features = preprocessing.scale(features)
scaled_features_latest_3 = scaled_features[-3:]
scaled_features = scaled_features[:-3]
data_frame.dropna(inplace=True)
label = np.array(data_frame['Real Price Label'])
```

Now, we can create a polynomial regression of degree **3** using the **fit_transform** method of **PolynomialFeatures**:

```
poly_regressor = PolynomialFeatures(degree=3)
poly_scaled_features = poly_regressor.fit_transform(scaled_features)
poly_scaled_features
```

The output of **poly_scaled_features** is as follows:

```
array([[ 1.        , -1.14839975, -1.13009904, ..., -1.52261953,
        -1.60632446, -1.69463102],
       [ 1.        , -1.14114523, -1.12483455, ..., -1.49346824,
        -1.56720585, -1.64458414],
       [ 1.        , -1.13389072, -1.12377394, ..., -1.48310475,
        -1.54991107, -1.61972667],
       ...,
       [ 1.        , -1.38866935,  2.97846643, ..., 31.70979016,
        38.05472653, 45.66924612],
       [ 1.        , -1.38866935,  2.83458633, ..., 29.05499915,
        37.06573938, 47.28511704],
       [ 1.        , -1.36429417,  2.95488131, ..., 31.89206605,
        39.42259303, 48.73126873]])
```

Then, we need to split the data into testing and training sets:

```
(poly_features_train, poly_features_test, \
poly_label_train, poly_label_test) = \
model_selection.train_test_split(poly_scaled_features, \
                                 label, test_size=0.1, \
                                 random_state=8)
```

The **train_test_split** function shuffles the lines of our data, keeps the correspondence, and puts approximately 10% of all data in the test variables, keeping 90% for the training variables. We also use **random_state=8** in order to reproduce the results.

Our data is now ready to be used for the multiple polynomial regression model; we will also measure its performance with the **score** function:

```
model = linear_model.LinearRegression()
model.fit(poly_features_train, poly_label_train)
model.score(poly_features_test, poly_label_test)
```

The output is as follows:

```
0.988000620369118
```

With a score or R^2 of **0.988**, our multiple polynomial regression model is slightly better than our multiple linear regression model (**0.985**), which we built in *Exercise 2.03*, *Preparing the Quandl Data for Prediction*. It might be possible that both models are overfitting the dataset.

There is another model in scikit-learn that performs polynomial regression, called the SVM model.

SUPPORT VECTOR REGRESSION

SVMs are binary classifiers and are usually used in classification problems (you will learn more about this in *Chapter 3*, *An Introduction to Classification*). An SVM classifier takes data and tries to predict which class it belongs to. Once the classification of a data point is determined, it gets labeled. But SVMs can also be used for regression; that is, instead of labeling data, it can predict future values in a series.

The SVR model uses the space between our data as a margin of error. Based on the margin of error, it makes predictions regarding future values.

If the margin of error is too small, we risk overfitting the existing dataset. If the margin of error is too big, we risk underfitting the existing dataset.

In the case of a classifier, the kernel describes the surface dividing the state space, whereas, in a regression, the kernel measures the margin of error. This kernel can use a linear model, a polynomial model, or many other possible models. The default kernel is **RBF**, which stands for **Radial Basis Function**.

SVR is an advanced topic that is outside the scope of this book. Therefore, we will only stick to an easy walk-through as an opportunity to try out another regression model on our data.

We can reuse the code from *Exercise 2.03, Preparing the Quandl Data for Prediction*, up to *Step 11*:

```
import quandl
import numpy as np
from sklearn import preprocessing
from sklearn import model_selection
from sklearn import linear_model
from matplotlib import pyplot as plot

data_frame = quandl.get("YALE/SPCOMP", \
                        start_date="1950-01-01", \
                        end_date="2019-12-31")

data_frame = data_frame[['Long Interest Rate', \
                         'Real Price', 'Real Dividend']]

data_frame.fillna(method='ffill', inplace=True)
data_frame['Real Price Label'] = data_frame['Real Price'].shift(-3)
features = np.array(data_frame.drop('Real Price Label', 1))
scaled_features = preprocessing.scale(features)
scaled_features_latest_3 = scaled_features[-3:]
scaled_features = scaled_features[:-3]
data_frame.dropna(inplace=True)
label = np.array(data_frame['Real Price Label'])

(features_train, features_test, label_train, label_test) = \
model_selection.train_test_split(scaled_features, label, \
                                 test_size=0.1, \
                                 random_state=8)
```

Then, we can perform a regression with **svm** by simply changing the linear model to a support vector model by using the **svm** method from **sklearn**:

```
from sklearn import svm
model = svm.SVR()
model.fit(features_train, label_train)
```

As you can see, performing an SVR is exactly the same as performing a linear regression, with the exception of defining the model as **svm.SVR()**.

Finally, we can predict and measure the performance of our model:

```
label_predicted = model.predict(features_test)
model.score(features_test, label_test)
```

The output is as follows:

```
0.03262153550014424
```

As you can see, the score or R^2 is quite low, our SVR's parameters need to be optimized in order to increase the accuracy of the model.

SUPPORT VECTOR MACHINES WITH A 3-DEGREE POLYNOMIAL KERNEL

Let's switch the kernel of the SVM to a polynomial function (the default degree is **3**) and measure the performance of the new model:

```
model = svm.SVR(kernel='poly')
model.fit(features_train, label_train)
label_predicted = model.predict(features_test)
model.score(features_test, label_test)
```

The output is as follows:

```
0.44465054598560627
```

We managed to increase the performance of the SVM by simply changing the kernel function to a polynomial function; however, the model still needs a lot of tuning to reach the same performance as the linear regression models.

ACTIVITY 2.01: BOSTON HOUSE PRICE PREDICTION WITH POLYNOMIAL REGRESSION OF DEGREES 1, 2, AND 3 ON MULTIPLE VARIABLES

In this activity, you will need to perform linear polynomial regression of degrees 1, 2, and 3 with scikit-learn and find the best model. You will work on the Boston House Prices dataset. The Boston House Price dataset is very famous and has been used as an example for research on regression models.

> **NOTE**
>
> More details about the Boston House Prices dataset can be found at https://archive.ics.uci.edu/ml/machine-learning-databases/housing/.
>
> The dataset file can also be found in our GitHub repository: https://packt.live/2V9kRUU.

You will need to predict the prices of houses in Boston (label) based on their characteristics (features). Your main goal will be to build 3 linear models using polynomial regressions of degrees **1**, **2**, and **3** with all the features of the dataset. You can find the following dataset description:

```
Boston house prices dataset
---------------------------

**Data Set Characteristics:**

    :Number of Instances: 506

    :Number of Attributes: 12 numeric/categorical predictive. Median Value (attribute 13 a) is
usually the target.

    :Attribute Information (in order):
        - CRIM     per capita crime rate by town
        - ZN       proportion of residential land zoned for lots over 25,000 sq.ft.
        - INDUS    proportion of non-retail business acres per town
        - CHAS     Charles River dummy variable (= 1 if tract bounds river; 0 otherwise)
        - NOX      nitric oxides concentration (parts per 10 million)
        - RM       average number of rooms per dwelling
        - AGE      proportion of owner-occupied units built prior to 1940
        - DIS      weighted distances to five Boston employment centres
        - RAD      index of accessibility to radial highways
        - TAX      full-value property-tax rate per $10,000
        - PTRATIO  pupil-teacher ratio by town
        - LSTAT    % lower status of the population
        - MEDV     Median value of owner-occupied homes in $1000's

    :Missing Attribute Values: None

    :Creator: Harrison, D. and Rubinfeld, D.L.
```

Figure 2.26: Boston housing dataset description

We will define our label as the **MEDV** field, which is the median value of the house in $1,000s. All of the other fields will be used as our features for our models. As this dataset does not contain any missing values, we won't have to replace missing values as we did in the previous exercises.

The following steps will help you to complete the activity:

1. Open a Jupyter Notebook.

2. Import the required packages and load the Boston House Prices data into a DataFrame.

3. Prepare the dataset for prediction by converting the label and features into NumPy arrays and scaling the features.

4. Create three different sets of features by transforming the scaled features into suitable formats for each of the polynomial regressions.

5. Split the data into training and testing sets with `random state = 8`.

6. Perform a polynomial regression of degree **1** and evaluate whether the model is overfitting.

7. Perform a polynomial regression of degree **2** and evaluate whether the model is overfitting.

8. Perform a polynomial regression of degree **3** and evaluate whether the model is overfitting.

9. Compare the predictions of the three models against the label on the testing set.

The expected output is this:

	label	model_1_prediction	model_2_prediction	model_3_prediction
0	18.5	19.269554	19.885620	21.067408
1	12.7	11.434612	14.470337	11.703696
2	21.9	37.610026	32.721497	-3713.860431
3	22.0	26.985628	26.337830	1.711389
4	50.0	40.548986	45.107178	-448.022868
5	36.2	27.730936	26.509888	45.594573
6	16.5	10.601832	18.077148	-6.973548
7	32.4	36.478511	32.267456	1584.038370
8	24.6	29.094356	27.081970	25.674404
9	50.0	34.409453	54.093201	-285.162605
10	13.9	14.138314	15.555481	14.549911
11	11.9	7.143616	14.295105	22.325523

Figure 2.27: Expected output based on the predictions

NOTE

The solution to this activity is available on page 334.

SUMMARY

In this chapter, we have learned the fundamentals of linear regression. After going through some basic mathematics, we looked at the mathematics of linear regression using one variable and multiple variables.

Then, we learned how to load external data from sources such as a CSV file, Yahoo Finance, and Quandl. After loading the data, we learned how to identify features and labels, how to scale data, and how to format data to perform regression.

We learned how to train and test a linear regression model, and how to predict the future. Our results were visualized by an easy-to-use Python graph plotting library called **pyplot**.

We also learned about a more complex form of linear regression: linear polynomial regression using arbitrary degrees. We learned how to define these regression problems on multiple variables and compare their performance on the Boston House Price dataset. As an alternative to polynomial regression, we also introduced SVMs as a regression model and experimented with two kernels.

In the next chapter, you will learn about classification and its models.

3

AN INTRODUCTION TO CLASSIFICATION

OVERVIEW

This chapter introduces you to classification. You will implement various techniques, such as k-nearest neighbors and SVMs. You will use the Euclidean and Manhattan distances to work with k-nearest neighbors. You will apply these concepts to solve intriguing problems such as predicting whether a credit card applicant has a risk of defaulting and determining whether an employee would stay with a company for more than two years. By the end of this chapter, you will be confident enough to work with any data using classification and come to a certain conclusion.

INTRODUCTION

In the previous chapter, you were introduced to regression models and learned how to fit a linear regression model with single or multiple variables, as well as with a higher-degree polynomial.

Unlike regression models, which focus on learning how to predict continuous numerical values (which can have an infinite number of values), classification, which will be introduced in this chapter, is all about splitting data into separate groups, also called classes.

For instance, a model can be trained to analyze emails and predict whether they are spam or not. In this case, the data is categorized into two possible groups (or classes). This type of classification is also called **binary classification**, which we will see a few examples of in this chapter. However, if there are more than two groups (or classes), you will be working on a **multi-class classification** (you will come across some examples of this in *Chapter 4*, *An Introduction to Decision Trees*).

But what is a real-world classification problem? Consider a model that tries to predict a given user's rating for a movie where this score can only take values: *like*, *neutral*, or *dislike*. This is a classification problem.

In this chapter, we will learn how to classify data using the k-nearest neighbors classifier and SVM algorithms. Just as we did for regression in the previous chapter, we will build a classifier based on cleaned and prepared training data and test the performance of our classifier using testing data.

We'll begin by looking at the fundamentals of classification.

THE FUNDAMENTALS OF CLASSIFICATION

As stated earlier, the goal of any classification problem is to separate the data into relevant groups accurately using a training set. There are a lot of applications of such projects in different industries, such as education, where a model can predict whether a student will pass or fail an exam, or healthcare, where a model can assess the level of severity of a given disease for each patient.

A classifier is a model that determines the label (output) or value (class) of any data point that it belongs to. For instance, suppose you have a set of observations that contains credit-worthy individuals, and another one that contains individuals that are risky in terms of their credit repayment tendencies.

Let's call the first group P and the second one Q. Here is an example of such data:

Income	Credit Cards	Mortgage	Savings	Class
88,000	4	-53,500	33,015	P
168,000	6	-337,000	50,312	P
23,000	1	-240,300	0	Q
34,000	2	-56,000	5,000	P
15,300	0	0	0	Q

Figure 3.1: Sample dataset

With this data, you will train a classification model that will be able to correctly classify a new observation into one of these two groups (this is binary classification). The model can find patterns such as a person with a salary above $60,000 being less risky or that having a mortgage/income ratio above ratio 10 makes an individual more at risk of not repaying their debts. This will be a **multi-class classification** exercise.

Classification models can be grouped into different families of algorithms. The most famous ones are as follows:

- Distance-based, such as **k-nearest neighbors**

- Linear models, such as **logistic regression** or **SVMs**

- Tree-based, such as **random forest**

In this chapter, you will be introduced to two algorithms from the first two types of family: k-nearest neighbors (distance-based) and SVMs (linear models).

> **NOTE**
>
> We'll walk you through tree-based algorithms such as random forest in *Chapter 4, An Introduction to Decision Trees*.

But before diving into the models, we need to clean and prepare the dataset that we will be using in this chapter.

In the following section, we will work on a German credit approvals dataset and perform all the data preparation required for the modeling stage. Let's start by loading the data.

EXERCISE 3.01: PREDICTING RISK OF CREDIT CARD DEFAULT (LOADING THE DATASET)

In this exercise, we will be loading a dataset into a pandas DataFrame and exploring its contents. We will use the dataset of German credit approvals to determine whether an individual presents a risk of defaulting.

> **NOTE**
>
> The CSV version of this dataset can be found on our GitHub repository:
>
> https://packt.live/3eriWTr.
>
> The original dataset and information regarding the dataset can be found at https://archive.ics.uci.edu/ml/datasets/Statlog+%28German+Credit+Data%29.
>
> The data files are located at https://archive.ics.uci.edu/ml/machine-learning-databases/statlog/german/.
>
> Citation - *Dua, D., & Graff, C.. (2017). UCI Machine Learning Repository*.

1. Open a new Jupyter Notebook file.

2. Import the **pandas** package as **pd**:

```
import pandas as pd
```

3. Create a new variable called **file_url**, which will contain the URL to the raw dataset file, as shown in the following code snippet:

```
file_url = 'https://raw.githubusercontent.com/'\
           'PacktWorkshops/'\
           'The-Applied-Artificial-Intelligence-Workshop/'\
           'master/Datasets/german_credit.csv'
```

4. Import the data using the **pd.read_csv()** method:

```
df = pd.read_csv(file_url)
```

5. Use **df.head()** to print the first five rows of the DataFrame:

```
df.head()
```

The expected output is this:

	default	account_check_status	duration_in_month	credit_history	purpose	credit_amount	savings	present_emp_since	installment_as_income_perc
0	0	< 0 DM	6	critical account/ other credits existing (not ...	domestic appliances	1169	unknown/ no savings account	.. >= 7 years	4
1	1	0 <= ... < 200 DM	48	existing credits paid back duly till now	domestic appliances	5951	... < 100 DM	1 <= ... < 4 years	2
2	0	no checking account	12	critical account/ other credits existing (not ...	(vacation - does not exist?)	2096	... < 100 DM	4 <= ... < 7 years	2
3	0	< 0 DM	42	existing credits paid back duly till now	radio/television	7882	... < 100 DM	4 <= ... < 7 years	2
4	1	< 0 DM	24	delay in paying off in the past	car (new)	4870	... < 100 DM	1 <= ... < 4 years	3

5 rows × 21 columns

Figure 3.2: The first five rows of the dataset

As you can see, the output in the preceding screenshot shows us the features of the dataset, which can be either numerical or categorical (text).

6. Now, use `df.tail()` to print the last five rows of the DataFrame:

```
df.tail()
```

The expected output is this:

	default	account_check_status	duration_in_month	credit_history	purpose	credit_amount	savings	present_emp_since	installment_as_income_perc
995	0	no checking account	12	existing credits paid back duly till now	radio/television	1736	... < 100 DM	4 <= ... < 7 years	3
996	0	< 0 DM	30	existing credits paid back duly till now	car (used)	3857	... < 100 DM	1 <= ... < 4 years	4
997	0	no checking account	12	existing credits paid back duly till now	domestic appliances	804	... < 100 DM	.. >= 7 years	4
998	1	< 0 DM	45	existing credits paid back duly till now	domestic appliances	1845	... < 100 DM	1 <= ... < 4 years	4
999	0	0 <= ... < 200 DM	45	critical account/ other credits existing (not ...	car (used)	4576	100 <= ... < 500 DM	unemployed	3

5 rows × 21 columns

Figure 3.3: The last five rows of the dataset

The last rows of the DataFrame are very similar to the first ones we saw earlier, so we can assume the structure is consistent across the rows.

7. Now, use **df.dtypes** to print the list of columns and their data types:

```
df.dtypes
```

The expected output is this:

```
default                        int64
account_check_status           object
duration_in_month              int64
credit_history                 object
purpose                        object
credit_amount                  int64
savings                        object
present_emp_since              object
installment_as_income_perc     int64
other_debtors                  object
present_res_since              int64
property                       object
age                            int64
other_installment_plans        object
housing                        object
credits_this_bank              int64
job                            object
people_under_maintenance       int64
telephone                      object
foreign_worker                 object
dtype: object
```

Figure 3.4: The list of columns and their data types

NOTE

To access the source code for this specific section, please refer to https://packt.live/3hQXJEs.

You can also run this example online at https://packt.live/3fN0DrT. You must execute the entire Notebook in order to get the desired result.

From the preceding output, we can see that this DataFrame has some numerical features (**int64**) but also text (**object**). We can also see that most of these features are either personal details for an individual, such as their age, or financial information such as credit history or credit amount.

By completing this exercise, we have successfully loaded the data into the DataFrame and had a first glimpse of the features and information it contains.

In the topics ahead, we will be looking at preprocessing this data.

DATA PREPROCESSING

Before building a classifier, we need to format our data so that we can keep relevant data in the most suitable format for classification and remove all the data that we are not interested in.

The following points are the best ways to achieve this:

- **Replacing or dropping values**:

  For instance, if there are **N/A** (or **NA**) values in the dataset, we may be better off substituting these values with a numeric value we can handle. Recall from the previous chapter that **NA** stands for **Not Available** and that it represents a missing value. We may choose to ignore rows with **NA** values or replace them with an outlier value.

  > **NOTE**
  >
  > An outlier value is a value such as -1,000,000 that clearly stands out from regular values in the dataset.

  The `fillna()` method of a DataFrame does this type of replacement. The replacement of **NA** values with an outlier looks as follows:

  ```
  df.fillna(-1000000, inplace=True)
  ```

  The `fillna()` method changes all **NA** values into numeric values.

  This numeric value should be far from any reasonable value in the DataFrame. Minus one million is recognized by the classifier as an exception, assuming that only positive values are there, as mentioned in the preceding note.

- **Dropping rows or columns**:

  The alternative to replacing missing values with extreme values is simply dropping these rows:

  ```
  df.dropna(0, inplace=True)
  ```

The first argument (value **0**) specifies that we drop rows, not columns. The second argument (**inplace=True**) specifies that we perform the drop operation without cloning the DataFrame, and will save the result in the same DataFrame. This DataFrame doesn't have any missing values, so the **dropna()** method didn't alter the DataFrame.

> **NOTE**
>
> Dropping the **NA** values is less desirable, as you often lose a reasonable chunk of your dataset.

If there is a column we do not want to include in the classification, we are better off dropping it. Otherwise, the classifier may detect false patterns in places where there is absolutely no correlation.

For instance, your phone number itself is very unlikely to correlate with your credit score. It is a 9 to 12-digit number that may very easily feed the classifier with a lot of noise. So, we can drop the **telephone** column, as shown in the following code snippet:

```
df.drop(['telephone'], 1, inplace=True)
```

The second argument (value **1**) indicates that we are dropping columns, instead of rows. The first argument is an enumeration of the columns we would like to drop (here, this is **['telephone']**). The **inplace** argument is used so that the call modifies the original DataFrame.

• **Transforming data**:

Often, the data format we are working with is not always optimal for the classification process. We may want to transform our data into a different format for multiple reasons, such as to highlight aspects of the data we are interested in (for example, Minmax scaling or normalization), to drop aspects of the data we are not interested in (for example, binarization), label encoding to transform categorical variables into numerical ones, and so on.

Minmax scaling scales each column in the data so that the lowest number in the column becomes 0, the highest number becomes 1, and all of the values in-between are proportionally scaled between 0 and 1.

This type of operation can be performed by the **MinMaxScaler** method of the scikit-learn **preprocessing** utility, as shown in the following code snippet:

```
from sklearn import preprocessing
import numpy as np

data = np.array([[19, 65], \
                 [4, 52], \
                 [2, 33]])

preprocessing.MinMaxScaler(feature_range=(0,1)).fit_transform(data)
```

The expected output is this:

```
array([[1.         , 1.         ],
       [0.11764706, 0.59375    ],
       [0.         , 0.         ]])
```

Binarization transforms data into ones and zeros based on a condition, as shown in the following code snippet:

```
preprocessing.Binarizer(threshold=10).transform(data)
```

The expected output is this:

```
array([[1, 1],
       [0, 1],
       [0, 1]])
```

In the preceding example, we transformed the original data (**[19, 65],[4, 52],[2, 33]**) into a binary form based on the condition of whether each value is greater than **10** or not (as defined by the **threshold=10** parameter). For instance, the first value, **19**, is above **10**, so it is replaced by **1** in the results.

Label encoding is important for preparing your features (inputs) for the modeling stage. While some of your features are string labels, scikit-learn algorithms expect this data to be transformed into numbers.

This is where the **preprocessing** library of scikit-learn comes into play.

> **NOTE**
>
> You might have noticed that in the credit scoring example, there were two data files. One contained labels in string form, while the other contained labels in integer form. We loaded the data with string labels so that you got some experience of how to preprocess data properly with the label encoder.

Label encoding is not rocket science. It creates a mapping between string labels and numeric values so that we can supply numbers to scikit-learn, as shown in the following example:

```
from sklearn import preprocessing
labels = ['Monday', 'Tuesday', 'Wednesday', \
          'Thursday', 'Friday']
label_encoder = preprocessing.LabelEncoder()
label_encoder.fit(labels)
```

Let's enumerate the encoding:

```
[x for x in enumerate(label_encoder.classes_)]
```

The expected output is this:

```
[(0, 'Friday'),
 (1, 'Monday'),
 (2, 'Thursday'),
 (3, 'Tuesday'),
 (4, 'Wednesday')]
```

The preceding result shows us that scikit-learn has created a mapping for each day of the week to a respective number; for example, **Friday** will be **0** and **Tuesday** will be **3**.

> **NOTE**
>
> By default, scikit-learn assigned the mapping number by sorting the original values alphabetically. This is why **Friday** is mapped to **0**.

Now, we can use this mapping (also called an encoder) to transform data.

Let's try this out on two examples, **Wednesday** and **Friday**, using the
transform() method:

```
label_encoder.transform(['Wednesday', 'Friday'])
```

The expected output is this:

```
array([4, 0], dtype=int64)
```

As expected, we got the results **4** and **0**, which are the mapping values for
Wednesday and **Friday**, respectively.

We can also use this encoder to perform the inverse transformation with the
inverse_transform function. Let's try this with the values **0** and **4**:

```
label_encoder.inverse_transform([0, 4])
```

The expected output is this:

```
array(['Friday', 'Wednesday'], dtype='<U9')
```

As expected, we got back the values **Friday** and **Wednesday**. Now, let's practice
what we've learned here on the German dataset.

EXERCISE 3.02: APPLYING LABEL ENCODING TO TRANSFORM CATEGORICAL VARIABLES INTO NUMERICAL VARIABLES

In this exercise, we will use one of the preprocessing techniques we just learned,
label encoding, to transform all categorical variables into numerical ones. This step is
necessary before training any machine learning model.

> **NOTE**
>
> We will be using the same dataset that we used in the previous exercise:
> the German credit approval dataset: https://packt.live/3eriWTr.

The following steps will help you complete this exercise:

1. Open a new Jupyter Notebook file.

2. Import the **pandas** package as **pd**:

    ```
    import pandas as pd
    ```

3. Create a new variable called **file_url**, which will contain the URL to the raw dataset:

```
file_url = 'https://raw.githubusercontent.com/'\
           'PacktWorkshops/'\
           'The-Applied-Artificial-Intelligence-Workshop/'\
           'master/Datasets/german_credit.csv'
```

4. Load the data using the **pd.read_csv()** method:

```
df = pd.read_csv(file_url)
```

5. Import **preprocessing** from **scikit-learn**:

```
from sklearn import preprocessing
```

6. Define a function called **fit_encoder()** that takes a DataFrame and a column name as parameters and will fit a label encoder on the values of the column. You will use **.LabelEncoder()** and **.fit()** from **preprocessing** and **.unique()** from **pandas** (this will extract all the possible values of a DataFrame column):

```
def fit_encoder(dataframe, column):
    encoder = preprocessing.LabelEncoder()
    encoder.fit(dataframe[column].unique())
    return encoder
```

7. Define a function called **encode()** that takes a DataFrame, a column name, and a label encoder as parameters and will transform the values of the column using the label encoder. You will use the **.transform()** method to do this:

```
def encode(dataframe, column, encoder):
    return encoder.transform(dataframe[column])
```

8. Create a new DataFrame called **cat_df** that contains only non-numeric columns and print its first five rows. You will use the **.select_dtypes()** method from pandas and specify **exclude='number'**:

```
cat_df = df.select_dtypes(exclude='number')
cat_df.head()
```

The expected output (not all columns are shown) is this:

	account_check_status	credit_history	purpose	savings	present_emp_since	other_debtors	property	other_installment_plans	housing
0	< 0 DM	critical account/ other credits existing (not ...	domestic appliances	unknown/ no savings account	.. >= 7 years	none	real estate	none	own
1	0 <= ... < 200 DM	existing credits paid back duly till now	domestic appliances	... < 100 DM	1 <= ... < 4 years	none	real estate	none	own
2	no checking account	critical account/ other credits existing (not ...	(vacation - does not exist?)	... < 100 DM	4 <= ... < 7 years	none	real estate	none	own
3	< 0 DM	existing credits paid back duly till now	radio/television	... < 100 DM	4 <= ... < 7 years	guarantor	if not A121 : building society savings agreeme...	none	for free
4	< 0 DM	delay in paying off in the past	car (new)	... < 100 DM	1 <= ... < 4 years	none	unknown / no property	none	for free

Figure 3.5: First five rows of the DataFrame containing only non-numeric columns

9. Create a list called **cat_cols** that contains the column name of **cat_df** and print its content. You will use **.columns** from pandas to do this:

```
cat_cols = cat_df.columns
cat_cols
```

The expected output is this:

```
Index(['account_check_status', 'credit_history', 'purpose',
       'savings', 'present_emp_since', 'other_debtors',
       'property', 'other_installment_plans', 'housing',
       'job', 'telephone', 'foreign_worker'], dtype='object')
```

10. Create a **for** loop that will iterate through each column from **cat_cols**, fit a label encoder using **fit_encoder()**, and transform the column with the **encode()** function:

```
for col in cat_cols:
    label_encoder = fit_encoder(df, col)
    df[col] = encode(df, col, label_encoder)
```

11. Print the first five rows of **df**:

```
df.head()
```

The expected output is this:

	default	account_check_status	duration_in_month	credit_history	purpose	credit_amount	savings	present_emp_since	installment_as_income_perc
0	0	1	6	1	4	1169	4	0	4
1	1	0	48	3	4	5951	1	2	2
2	0	3	12	1	0	2096	1	3	2
3	0	1	42	3	7	7882	1	3	2
4	1	1	24	2	2	4870	1	2	3

Figure 3.6: First five rows of the encoded DataFrame

> **NOTE**
>
> To access the source code for this specific section, please refer to
> https://packt.live/2Njh57h.
>
> You can also run this example online at https://packt.live/2YZhtx5. You must
> execute the entire Notebook in order to get the desired result.

We have successfully encoded non-numeric columns. Now, our DataFrame contains
only numeric values.

IDENTIFYING FEATURES AND LABELS

Before training our model, we still have to perform two final steps. The first one
is to separate our features from the label (also known as a response variable or
dependent variable). The **label** column is the one we want our model to predict. For
the German credit dataset, in our case, it will be the column called **default**, which
tells us whether an individual will present a risk of defaulting or not.

The features are all the other columns present in the dataset. The model will use the
information contained in those columns and find the relevant patterns in order to
accurately predict the corresponding label.

The scikit-learn package requires the labels and features to be stored in two different
variables. Luckily, the pandas package provides a method to extract a column from a
DataFrame called **.pop()**.

We will extract the **default** column and store it in a variable called **label**:

```
label = df.pop('default')
label
```

The expected output is this:

```
0       0
1       1
2       0
3       0
4       1
       ..
995     0
996     0
997     0
998     1
999     0
Name: default, Length: 1000, dtype: int64
```

Now, if we look at the content of **df**, we will see that the **default** column is not present anymore:

```
df.columns
```

The expected output is this:

```
Index(['account_check_status', 'duration_in_month',
       'credit_history', 'purpose', 'credit_amount',
       'savings', 'present_emp_since',
       'installment_as_income_perc', 'other_debtors',
       'present_res_since', 'property', 'age',
       'other_installment_plans', 'housing',
       'credits_this_bank', 'job', 'people_under_maintenance',
       'telephone', 'foreign_worker'],
      dtype='object')
```

Now that we have our features and labels ready, we need to split our dataset into training and testing sets.

SPLITTING DATA INTO TRAINING AND TESTING USING SCIKIT-LEARN

The final step that's required before training a classifier is to split our data into training and testing sets. We already saw how to do this in *Chapter 2, An Introduction to Regression*:

```
from sklearn import model_selection

features_train, features_test, \
label_train, label_test = \
model_selection.train_test_split(df, label, test_size=0.1, \
                                 random_state=8)
```

The **train_test_split** method shuffles and then splits our features and labels into a training dataset and a testing dataset.

We can specify the size of the testing dataset as a number between **0** and **1**. A **test_size** of **0.1** means that **10%** of the data will go into the testing dataset. You can also specify a **random_state** so that you get the exact same split if you run this code again.

We will use the training set to train our classifier and use the testing set to evaluate its predictive performance. By doing so, we can assess whether our model is overfitting and has learned patterns that are only relevant to the training set.

In the next section, we will introduce you to the famous k-nearest neighbors classifier.

THE K-NEAREST NEIGHBORS CLASSIFIER

Now that we have our training and testing data, it is time to prepare our classifier to perform k-nearest neighbor classification. After being introduced to the k-nearest neighbor algorithm, we will use scikit-learn to perform classification.

INTRODUCING THE K-NEAREST NEIGHBORS ALGORITHM (KNN)

The goal of classification algorithms is to divide data so that we can determine which data points belong to which group.

Suppose that a set of classified points is given to us. Our task is to determine which class a new data point belongs to.

In order to train a k-nearest neighbor classifier (also referred to as KNN), we need to provide the corresponding class for each observation on the training set, that is, which group it belongs to. The goal of the algorithm is to find the relevant relationship or patterns between the features that will lead to this class. The k-nearest neighbors algorithm is based on a proximity measure that calculates the distance between data points.

The two most famous proximity (or distance) measures are the Euclidean and the Manhattan distance. We will go through more details in the next section.

For any new given point, KNN will find its k nearest neighbor, see which class is the most frequent between those k neighbors, and assign it to this new observation. But what is k, you may ask? Determining the value of k is totally arbitrary. You will have to set this value upfront. This is not a parameter that can be learned by the algorithm; it needs to be set by data scientists. This kind of parameter is called a **hyperparameter**. Theoretically, you can set the value of k to between 1 and positive infinity.

There are two main best practices to take into consideration:

- k should always be an odd number. The reason behind this is that we want to avoid a situation that ends in a tie. For instance, if you set *k=4* and it so happens that two of the neighbors of a point are from class A and the other two are from class B, then KNN doesn't know which class to choose. To avoid this situation, it is better to choose *k=3* or *k=5*.

- The greater k is, the more accurate KNN will be. For example, if we compare the cases between *k=1* and *k=15*, the second one will give you more confidence that KNN will choose the right class as it will need to look at more neighbors before making a decision. On the other hand, with *k=1*, it only looks at the closest neighbor and assigns the same class to an observation. But how can we be sure it is not an outlier or a special case? Asking more neighbors will lower the risk of making the wrong decision. But there is a drawback to this: the higher k is, the longer it will take KNN to make a prediction. This is because it will have to perform more calculations to get the distance between all the neighbors of an observation. Due to this, you have to find the sweet spot that will give correct predictions without compromising too much on the time it takes to make a prediction.

DISTANCE METRICS WITH K-NEAREST NEIGHBORS CLASSIFIER IN SCIKIT-LEARN

Many distance metrics could work with the k-nearest neighbors algorithm. We will present the two most frequently used ones: the Euclidean distance and the Manhattan distance of two data points.

THE EUCLIDEAN DISTANCE

The distance between two points, **A** and **B**, with the coordinates **A=(a1, a2, …, an)** and **B=(b1, b2, …, bn)**, respectively, is the length of the line connecting these two points. For example, if A and B are two-dimensional data points, the Euclidean distance, **d**, will be as follows:

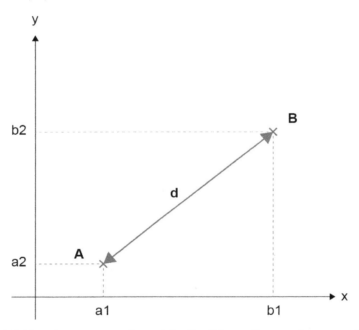

Figure 3.7: Visual representation of the Euclidean distance between A and B

The formula to calculate the Euclidean distance is as follows:

$$\text{distance}(a,b) = \sqrt{\sum_{i=1}^{n} (a_i - b_i)^2}$$

Figure 3.8: Distance between points A and B

As we will be using the Euclidean distance in this book, let's see how we can use scikit-learn to calculate the distance of multiple points.

We have to import **euclidean_distances** from **sklearn.metrics. pairwise**. This function accepts two sets of points and returns a matrix that contains the pairwise distance of each point from the first and second sets of points.

Let's take the example of an observation, Z, with coordinates (**4 , 4**). Here, we want to calculate the Euclidean distance with 3 others points, A, B, and C, with the coordinates (**2 , 3**), (**3 , 7**), and (**1 , 6**), respectively:

```
from sklearn.metrics.pairwise import euclidean_distances

observation = [4,4]
neighbors = [[2,3], [3,7], [1,6]]

euclidean_distances([observation], neighbors)
```

The expected output is this:

```
array([[2.23606798, 3.16227766, 3.60555128]])
```

Here, the distance of Z=(**4 , 4**) and B=(**3 , 7**) is approximately **3.162**, which is what we got in the output.

We can also calculate the Euclidean distances between points in the same set:

```
euclidean_distances(neighbors)
```

The expected output is this:

```
array([[0.        , 4.12310563, 3.16227766],
       [4.12310563, 0.        , 2.23606798],
       [3.16227766, 2.23606798, 0.        ]])
```

The diagonal that contains value **0** corresponds to the Euclidean distance between each data point and itself. This matrix is symmetric from this diagonal as it calculates the distance of two points and its reverse. For example, the value **4.12310563** on the first row is the distance between A and B, while the same value on the second row corresponds to the distance between B and A.

THE MANHATTAN/HAMMING DISTANCE

The formula of the Manhattan (or Hamming) distance is very similar to the Euclidean distance, but rather than using the square root, it relies on calculating the absolute value of the difference of the coordinates of the data points:

$$\text{distance}(a,b) = \sum_{i=1}^{n} \left| a_i - b_i \right|$$

Figure 3.9: The Manhattan and Hamming distance

You can think of the Manhattan distance as if we're using a grid to calculate the distance rather than using a straight line:

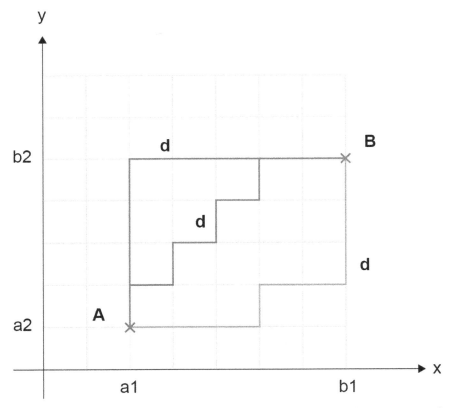

Figure 3.10: Visual representation of the Manhattan distance between A and B

As shown in the preceding plot, the Manhattan distance will follow the path defined by the grid to point B from A.

Another interesting property is that there can be multiple shortest paths between A and B, but their Manhattan distances will all be equal to each other. In the preceding example, if each cell of the grid equals a unit of 1, then all three of the shortest paths highlighted will have a Manhattan distance of 9.

The Euclidean distance is a more accurate generalization of distance, while the Manhattan distance is slightly easier to calculate as you only need to find the difference between the absolute value rather than calculating the difference between squares and then taking the root.

EXERCISE 3.03: ILLUSTRATING THE K-NEAREST NEIGHBORS CLASSIFIER ALGORITHM IN MATPLOTLIB

Suppose we have a list of employee data. Our features are the number of hours worked per week and the yearly salary. Our label indicates whether an employee has stayed with our company for more than 2 years. The length of stay is represented by zero if it is less than 2 years and one if it is greater than or equal to 2 years.

We want to create a three-nearest neighbors classifier that determines whether an employee will stay with our company for at least 2 years.

Then, we would like to use this classifier to predict whether an employee with a request to work 32 hours a week and earning 52,000 dollars per year is going to stay with the company for 2 years or not.

Follow these steps to complete this exercise:

> **NOTE**
>
> The aforementioned dataset is available on GitHub at
> https://packt.live/2V5VaV9.

1. Open a new Jupyter Notebook file.

2. Import the **pandas** package as **pd**:

```
import pandas as pd
```

3. Create a new variable called **file_url()**, which will contain the URL to the raw dataset:

```
file_url = 'https://raw.githubusercontent.com/'\
            'PacktWorkshops/'\
            'The-Applied-Artificial-Intelligence-Workshop/'\
            'master/Datasets/employees_churned.csv'
```

4. Load the data using the **pd.read_csv()** method:

```
df = pd.read_csv(file_url)
```

5. Print the rows of the DataFrame:

```
df
```

The expected output is this:

	hours_worked	salary	over_two_years
0	20	50000	0
1	24	45000	0
2	32	48000	0
3	24	55000	0
4	40	50000	0
5	40	62000	1
6	40	48000	1
7	32	55000	1
8	40	72000	1
9	32	60000	1

Figure 3.11: DataFrame of the employees dataset

6. Import **preprocessing** from **scikit-learn**:

```
from sklearn import preprocessing
```

7. Instantiate a **MinMaxScaler** with **feature_range=(0,1)** and save it to a variable called **scaler**:

```
scaler = preprocessing.MinMaxScaler(feature_range=(0,1))
```

8. Scale the DataFrame using `.fit_transform()`, save the results in a new variable called **scaled_employees**, and print its content:

```
scaled_employees = scaler.fit_transform(df)
scaled_employees
```

The expected output is this:

```
array([[0.       , 0.18518519, 0.       ],
       [0.2      , 0.       , 0.       ],
       [0.6      , 0.11111111, 0.       ],
       [0.2      , 0.37037037, 0.       ],
       [1.       , 0.18518519, 0.       ],
       [1.       , 0.62962963, 1.       ],
       [1.       , 0.11111111, 1.       ],
       [0.6      , 0.37037037, 1.       ],
       [1.       , 1.       , 1.       ],
       [0.6      , 0.55555556, 1.       ]])
```

In the preceding code snippet, we have scaled our original dataset so that all the values range between 0 and 1.

9. From the scaled data, extract each of the three columns and save them into three variables called **hours_worked**, **salary**, and **over_two_years**, as shown in the following code snippet:

```
hours_worked = scaled_employees[:, 0]
salary = scaled_employees[:, 1]
over_two_years = scaled_employees[:, 2]
```

10. Import the **matplotlib.pyplot** package as **plt**:

```
import matplotlib.pyplot as plt
```

11. Create two scatter plots with **plt.scatter** using **hours_worked** as the x-axis and **salary** as the y-axis, and then create different markers according to the value of **over_two_years**. You can add the labels for the x and y axes with **plt.xlabel** and **plt.ylabel**. Display the scatter plots with **plt.show()**:

```
plt.scatter(hours_worked[:5], salary[:5], marker='+')
plt.scatter(hours_worked[5:], salary[5:], marker='o')
plt.xlabel("hours_worked")
plt.ylabel("salary")
plt.show()
```

The expected output is this:

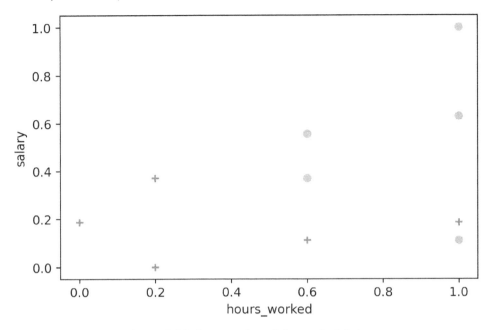

Figure 3.12: Scatter plot of the scaled data

In the preceding code snippet, we have displayed the data points of the scaled data on a scatter plot. The **+** points represent the employees that stayed less than 2 years, while the **o** ones are for the employees who stayed for more than 2 years.

Now, let's say we got a new observation and we want to calculate the Euclidean distance with the data from the scaled dataset.

12. Create a new variable called **observation** with the coordinates **[0.5, 0.26]**:

```
observation = [0.5, 0.26]
```

13. Import the **euclidean_distances** function from **sklearn. metrics.pairwise**:

```
from sklearn.metrics.pairwise import euclidean_distances
```

14. Create a new variable called **features**, which will extract the first two columns of the scaled dataset:

```
features = scaled_employees[:,:2]
```

15. Calculate the Euclidean distance between **observation** and **features** using **euclidean_distances**, save it into a variable called **dist**, and print its value, as shown in the following code snippet:

```
dist = euclidean_distances([observation], features)
dist
```

The expected output is this:

```
array([[0.50556627, 0.39698866, 0.17935412, 0.3196586 ,
        0.50556627, 0.62179262, 0.52169714, 0.14893495,
        0.89308454, 0.31201456]])
```

> **NOTE**
>
> To access the source code for this specific section, please refer to
> https://packt.live/3djY1jO.
>
> You can also run this example online at https://packt.live/3esx7HF. You must
> execute the entire Notebook in order to get the desired result.

From the preceding output, we can see that the three nearest neighbors are as follows:

- **0.1564897** for point **[0.6, 0.37037037, 1.]**

- **0.17114358** for point **[0.6, 0.11111111, 0.]**

- **0.32150303** for point **[0.6, 0.55555556, 1.]**

If we choose **k=3**, KNN will look at the classes for these three nearest neighbors and since two of them have a label of **1**, it will assign this class to our new observation, **[0.5, 0.26]**. This means that our three-nearest neighbors classifier will classify this new employee as being more likely to stay for at least 2 years.

By completing this exercise, we saw how a KNN classifier will classify a new observation by finding its three closest neighbors using the Euclidean distance and then assign the most frequent class to it.

PARAMETERIZATION OF THE K-NEAREST NEIGHBORS CLASSIFIER IN SCIKIT-LEARN

The parameterization of the classifier is where you fine-tune the accuracy of your classifier. Since we haven't learned all of the possible variations of k-nearest neighbors, we will concentrate on the parameters that you will understand based on this topic:

> **NOTE**
>
> You can access the documentation of the k-nearest neighbors classifier here: http://scikit-learn.org/stable/modules/generated/sklearn.neighbors.KNeighborsClassifier.html.

- **n_neighbors**: This is the k value of the k-nearest neighbors algorithm. The default value is **5**.

- **metric**: When creating the classifier, you will see a name – **Minkowski**. Don't worry about this name – you have learned about the first- and second-order Minkowski metrics already. This metric has a **power** parameter. For **p=1**, the Minkowski metric is the same as the Manhattan metric. For **p=2**, the Minkowski metric is the same as the Euclidean metric.

- **p**: This is the power of the Minkowski metric. The default value is **2**.

You have to specify these parameters once you create the classifier:

```
classifier = neighbors.KNeighborsClassifier(n_neighbors=50, p=2)
```

Then, you will have to fit the KNN classifier with your training data:

```
classifier.fit(features, label)
```

The **predict()** method can be used to predict the label for any new data point:

```
classifier.predict(new_data_point)
```

In the next exercise, we will be using the KNN implementation from scikit-learn to automatically find the nearest neighbors and assign corresponding classes.

EXERCISE 3.04: K-NEAREST NEIGHBORS CLASSIFICATION IN SCIKIT-LEARN

In this exercise, we will use scikit-learn to automatically train a KNN classifier on the German credit approval dataset and try out different values for the **n_neighbors** and **p** hyperparameters to get the optimal output values. We will need to scale the data before fitting KNN.

Follow these steps to complete this exercise:

> **NOTE**
>
> This exercise is a follow up from *Exercise 3.02*, *Applying Label Encoding to Transform Categorical Variables into Numerical*. We already saved the resulting dataset from *Exercise 3.02*, *Applying Label Encoding to Transform Categorical Variables into Numerical* in the GitHub repository at https://packt.live/2Yqdb2Q.

1. Open a new Jupyter Notebook.

2. Import the **pandas** package as **pd**:

```
import pandas as pd
```

3. Create a new variable called **file_url**, which will contain the URL to the raw dataset:

```
file_url = 'https://raw.githubusercontent.com/'\
           'PacktWorkshops/'\
           'The-Applied-Artificial-Intelligence-Workshop/'\
           'master/Datasets/german_prepared.csv'
```

4. Load the data using the **pd.read_csv()** method:

```
df = pd.read_csv(file_url)
```

5. Import **preprocessing** from **scikit-learn**:

```
from sklearn import preprocessing
```

6. Instantiate **MinMaxScaler** with **feature_range=(0,1)** and save it to a variable called **scaler**:

```
scaler = preprocessing.MinMaxScaler(feature_range=(0,1))
```

7. Fit the scaler and apply the corresponding transformation to the DataFrame using `.fit_transform()` and save the results to a variable called **scaled_credit**:

```
scaled_credit = scaler.fit_transform(df)
```

8. Extract the **response** variable (the first column) to a new variable called **label**:

```
label = scaled_credit[:, 0]
```

9. Extract the features (all the columns except for the first one) to a new variable called **features**:

```
features = scaled_credit[:, 1:]
```

10. Import **model_selection.train_test_split** from **sklearn**:

```
from sklearn.model_selection import train_test_split
```

11. Split the scaled dataset into training and testing sets with **test_size=0.2** and **random_state=7** using **train_test_split**:

```
features_train, features_test, \
label_train, label_test = \
train_test_split(features, label, test_size=0.2, \
                 random_state=7)
```

12. Import **neighbors** from **sklearn**:

```
from sklearn import neighbors
```

13. Instantiate **KNeighborsClassifier** and save it to a variable called **classifier**:

```
classifier = neighbors.KNeighborsClassifier()
```

14. Fit the k-nearest neighbors classifier on the training set:

```
classifier.fit(features_train, label_train)
```

Since we have not mentioned the value of k, the default is **5**.

15. Print the accuracy score for the training set with `.score()`:

```
acc_train = classifier.score(features_train, label_train)
acc_train
```

You should get the following output:

```
0.78625
```

With this, we've achieved an accuracy score of **0.78625** on the training set with the default hyperparameter values: *k=5* and the Euclidean distance.

Let's have a look at the score for the testing set.

16. Print the accuracy score for the testing set with **.score()**:

```
acc_test = classifier.score(features_test, label_test)
acc_test
```

You should get the following output:

```
0.75
```

The accuracy score dropped to **0.75** on the testing set. This means our model is overfitting and doesn't generalize well to unseen data. In the next activity, we will try different hyperparameter values and see if we can improve this.

> **NOTE**
>
> To access the source code for this specific section, please refer to
> https://packt.live/2ATeluO.
>
> You can also run this example online at https://packt.live/2VbDTKx. You must execute the entire Notebook in order to get the desired result.

In this exercise, we learned how to split a dataset into training and testing sets and fit a KNN algorithm. Our final model can accurately predict whether an individual is more likely to default or not 75% of the time.

ACTIVITY 3.01: INCREASING THE ACCURACY OF CREDIT SCORING

In this activity, you will be implementing the parameterization of the k-nearest neighbors classifier and observing the end result. The accuracy of credit scoring is currently 75%. You need to find a way to increase it by a few percentage points.

You can try different values for k (**5**, **10**, **15**, **25**, and **50**) with the Euclidean and Manhattan distances.

> **NOTE**
>
> This activity requires you to complete *Exercise 3.04, K-Nearest Neighbors Classification in scikit-learn* first as we will be using the previously prepared data here.

The following steps will help you complete this activity:

1. Import **neighbors** from **sklearn**.

2. Create a function to instantiate **KNeighborsClassifier** with hyperparameters specified, fit it with the training data, and return the accuracy score for the training and testing sets.

3. Using the function you created, assess the accuracy score for k = (**5**, **10**, **15**, **25**, **50**) for both the Euclidean and Manhattan distances.

4. Find the best combination of hyperparameters.

The expected output is this:

```
(0.775, 0.785)
```

> **NOTE**
>
> The solution to this activity can be found on page 343.

In the next section, we will introduce you to another machine learning classifier: a **Support Vector Machine (SVM)**.

CLASSIFICATION WITH SUPPORT VECTOR MACHINES

We first used SVMs for regression in *Chapter 2, An Introduction to Regression*. In this topic, you will find out how to use SVMs for classification. As always, we will use scikit-learn to run our examples in practice.

WHAT ARE SUPPORT VECTOR MACHINE CLASSIFIERS?

The goal of an SVM is to find a surface in an n-dimensional space that separates the data points in that space into multiple classes.

In two dimensions, this surface is often a straight line. However, in three dimensions, the SVM often finds a plane. These surfaces are optimal in the sense that they are based on the information available to the machine so that it can optimize the separation of the n-dimensional spaces.

The optimal separator found by the SVM is called the best separating hyperplane.

An SVM is used to find one surface that separates two sets of data points. In other words, SVMs are **binary classifiers**. This does not mean that SVMs can only be used for binary classification. Although we were only talking about one plane, SVMs can be used to partition a space into any number of classes by generalizing the task itself.

The separator surface is optimal in the sense that it maximizes the distance of each data point from the separator surface.

A vector is a mathematical structure defined on an n-dimensional space that has a magnitude (length) and a direction. In two dimensions, you draw the vector (x, y) from the origin to the point (x, y). Based on geometry, you can calculate the length of the vector using the Pythagorean theorem and the direction of the vector by calculating the angle between the horizontal axis and the vector.

For instance, in two dimensions, the vector (3, -4) has the following magnitude:

```
np.sqrt( 3 * 3 + 4 * 4 )
```

The expected output is this:

```
5.0
```

It has the following direction (in degrees):

```
np.arctan(-4/3) / 2 / np.pi * 360
```

The expected output is this:

```
-53.13010235415597
```

UNDERSTANDING SUPPORT VECTOR MACHINES

Suppose that two sets of points with two different classes, 0 and 1, are given. For simplicity, we can imagine a two-dimensional plane with two features: one mapped on the horizontal axis and one mapped on the vertical axis.

The objective of the SVM is to find the best separating line that separates points **A**, **D**, **C**, **B**, and **H**, which all belong to class 0, from points **E**, **F**, and **G**, which are of class 1:

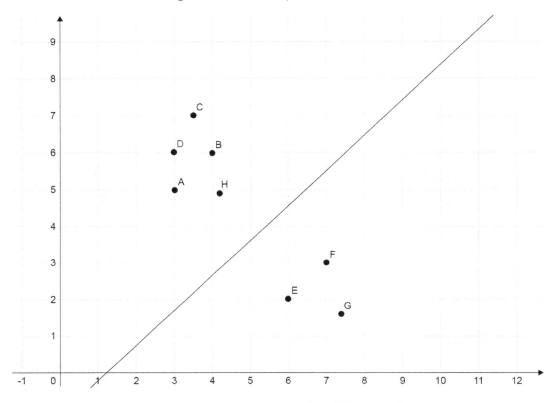

Figure 3.13: Line separating red and blue members

But separation is not always that obvious. For instance, if there is a new point of class 0 in-between **E**, **F**, and **G**, there is no line that could separate all the points without causing errors. If the points from class 0 form a full circle around the class 1 points, there is no straight line that could separate the two sets:

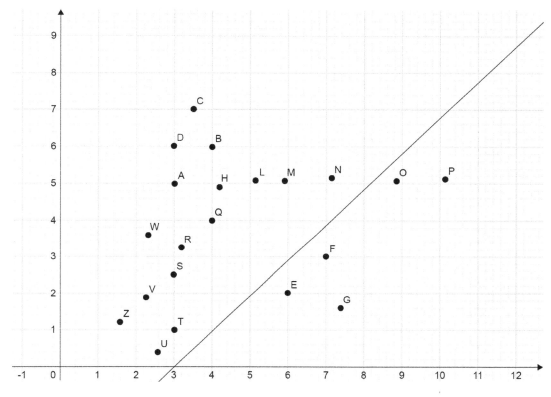

Figure 3.14: Graph with two outlier points

For instance, in the preceding graph, we tolerate two outlier points, **O** and **P**.

In the following solution, we do not tolerate outliers, and instead of a line, we create the best separating path consisting of two half-lines:

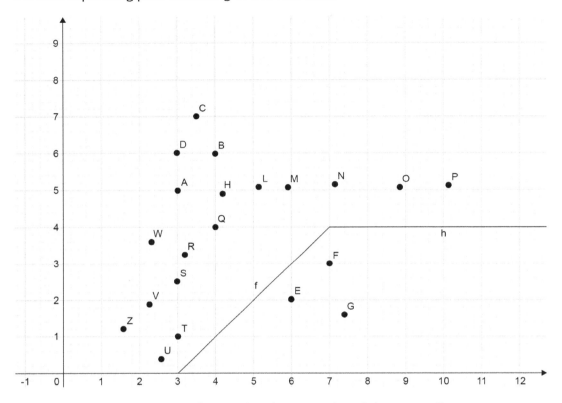

Figure 3.15: Graph removing the separation of the two outliers

The perfect separation of all data points is rarely worth the resources. Therefore, the SVM can be regularized to simplify and restrict the definition of the best separating shape and allow outliers.

The regularization parameter of an SVM determines the rate of errors to allow or forbid misclassifications.

An SVM has a kernel parameter. A linear kernel strictly uses a linear equation to describe the best separating hyperplane. A polynomial kernel uses a polynomial, while an exponential kernel uses an exponential expression to describe the hyperplane.

A margin is an area centered around the separator and is bounded by the points closest to the separator. A balanced margin has points from each class that are equidistant from the line.

When it comes to defining the allowed error rate of the best separating hyperplane, a gamma parameter decides whether only the points near the separator count in determining the position of the separator, or whether the points farthest from the line count, too. The higher the gamma, the lower the number of points that influence the location of the separator.

SUPPORT VECTOR MACHINES IN SCIKIT-LEARN

Our entry point is the end result of *Activity 3.02, Support Vector Machine Optimization in scikit-learn*. Once we have split the training and test data, we are ready to set up the classifier:

```
features_train, features_test, \
label_train, label_test = \
model_selection.train_test_split(scaled_features, label,\
                                 test_size=0.2)
```

Instead of using the k-nearest neighbors classifier, we will use the **svm. SVC()** classifier:

```
from sklearn import svm
classifier = svm.SVC()
classifier.fit(features_train, label_train)
classifier.score(features_test, label_test)
```

The expected output is this:

```
0.745
```

It seems that the default SVM classifier of scikit-learn does a slightly better job than the k-nearest neighbors classifier.

PARAMETERS OF THE SCIKIT-LEARN SVM

The following are the parameters of the scikit-learn SVM:

- **kernel**: This is a string or callable parameter specifying the kernel that's being used in the algorithm. The predefined kernels are **linear**, **poly**, **rbf**, **sigmoid**, and **precomputed**. The default value is **rbf**.

- **degree**: When using a polynomial, you can specify the degree of the polynomial. The default value is **3**.

- **gamma**: This is the kernel coefficient for **rbf**, **poly**, and **sigmoid**. The default value is **auto**, which is computed as *1/number_of_features*.

- **C**: This is a floating-point number with a default of **1.0** that describes the penalty parameter of the error term.

> **NOTE**
>
> You can read about the parameters in the reference documentation at
> http://scikit-learn.org/stable/modules/generated/sklearn.svm.SVC.html.

Here is an example of an SVM:

```
classifier = svm.SVC(kernel="poly", C=2, degree=4, gamma=0.05)
```

ACTIVITY 3.02: SUPPORT VECTOR MACHINE OPTIMIZATION IN SCIKIT-LEARN

In this activity, you will be using, comparing, and contrasting the different SVMs' classifier parameters. With this, you will find a set of parameters resulting in the highest classification data on the training and testing data that we loaded and prepared in *Activity 3.01, Increasing the Accuracy of Credit Scoring*.

You must different combinations of hyperparameters for SVM:

- **kernel="linear"**
- **kernel="poly"**, **C=1, degree=4, gamma=0.05**
- **kernel="poly"**, **C=1, degree=4, gamma=0.05**
- **kernel="poly"**, **C=1, degree=4, gamma=0.25**
- **kernel="poly"**, **C=1, degree=4, gamma=0.5**
- **kernel="poly"**, **C=1, degree=4, gamma=0.16**
- **kernel="sigmoid"**
- **kernel="rbf"**, **gamma=0.15**
- **kernel="rbf"**, **gamma=0.25**
- **kernel="rbf"**, **gamma=0.5**
- **kernel="rbf"**, **gamma=0.35**

The following steps will help you complete this activity:

1. Open a new Jupyter Notebook file and execute all the steps mentioned in the previous, *Exercise 3.04, K-Nearest Neighbor Classification in scikit-learn.*

2. Import **svm** from **sklearn**.

3. Create a function to instantiate an SVC with the hyperparameters specified, fit with the training data, and return the accuracy score for the training and testing sets.

4. Using the function you created, assess the accuracy scores for the different hyperparameter combinations.

5. Find the best combination of hyperparameters.

The expected output is this:

```
(0.78125, 0.775)
```

> **NOTE**
>
> The solution for this activity can be found on page 347.

SUMMARY

In this chapter, we learned about the basics of classification and the difference between regression problems. Classification is about predicting a response variable with limited possible values. As for any data science project, data scientists need to prepare the data before training a model. In this chapter, we learned how to standardize numerical values and replace missing values. Then, you were introduced to the famous k-nearest neighbors algorithm and discovered how it uses distance metrics to find the closest neighbors to a data point and then assigns the most frequent class among them. We also learned how to apply an SVM to a classification problem and tune some of its hyperparameters to improve the performance of the model and reduce overfitting.

In the next chapter, we will walk you through a different type of algorithm, called decision trees.

4

AN INTRODUCTION TO DECISION TREES

OVERVIEW

This chapter introduces you to two types of supervised learning algorithms in detail. The first algorithm will help you classify data points using decision trees, while the other algorithm will help you classify data points using random forests. Furthermore, you'll learn how to calculate the precision, recall, and F_1 score of models, both manually and automatically. By the end of this chapter, you will be able to analyze the metrics that are used for evaluating the utility of a data model and classify data points based on decision trees and random forest algorithms.

INTRODUCTION

In the previous two chapters, we learned the difference between regression and classification problems, and we saw how to train some of the most famous algorithms. In this chapter, we will look at another type of algorithm: tree-based models.

Tree-based models are very popular as they can model complex non-linear patterns and they are relatively easy to interpret. In this chapter, we will introduce you to decision trees and the random forest algorithms, which are some of the most widely used tree-based models in the industry.

DECISION TREES

A decision tree has leaves, branches, and nodes. Nodes are where a decision is made. A decision tree consists of rules that we use to formulate a decision (or prediction) on the prediction of a data point.

Every node of the decision tree represents a feature, while every edge coming out of an internal node represents a possible value or a possible interval of values of the tree. Each leaf of the tree represents a label value of the tree.

This may sound complicated, but let's look at an application of this.

Suppose we have a dataset with the following features and the response variable is determining whether a person is creditworthy or not:

ID	Employed	Income	LoanType	LoanAmount	CreditWorthy
A	false	75,000	car	30,000	No
B	false	25,000	studies	15,000	No
C	true	125,000	car	30,000	Yes
D	true	75,000	car	30,000	Yes
E	true	100,000	studies	25,000	Yes
F	true	100000	house	125000	Yes
G	false	80000	house	150000	Yes

Figure 4.1: Sample dataset to formulate the rules

A decision tree, remember, is just a group of rules. Looking at the dataset in *Figure 4.1*, we can come up with the following rules:

- All people with house loans are determined as creditworthy.

- If debtors are employed and studying, then loans are creditworthy.

- People with income above 75,000 a year are creditworthy.

- At or below 75,000 a year, people with car loans and who are employed are creditworthy.

Following the order of the rules we just defined, we can build a tree, as shown in *Figure 4.2* and describe one possible credit scoring method:

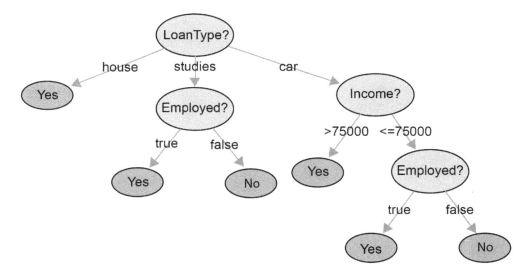

Figure 4.2: Decision tree for the loan type

First, we determine the loan type. House loans are automatically creditworthy according to the first rule. Study loans are described by the second rule, resulting in a subtree containing another decision on employment. Since we have covered both house and study loans, there are only car loans left. The third rule describes an income decision, while the fourth rule describes a decision on employment.

Whenever we must score a new debtor to determine whether they are creditworthy, we have to go through the decision tree from top to bottom and observe the true or false value at the bottom.

Obviously, a model based on seven data points is highly inaccurate because we can't generalize rules that simply do not match reality. Therefore, rules are often determined based on large amounts of data.

This is not the only way that we can create a decision tree. We can build decision trees based on other sequences of rules, too. Let's extract some other rules from the dataset in *Figure 4.1*.

Observation 1: Notice that individual salaries that are greater than 75,000 are all creditworthy.

Rule 1: `Income > 75,000 => CreditWorthy` is true.

Rule 1 classifies four out of seven data points (IDs C, E, F, G); we need more rules for the remaining three data points.

Observation 2: Out of the remaining three data points, two are not employed. One is employed (ID D) and is creditworthy. With this, we can claim the following rule:

Rule 2: Assuming `Income <= 75,000`, the following holds true: `Employed == true => CreditWorthy`.

Note that with this second rule, we can also classify the remaining two data points (IDs A and B) as not creditworthy. With just two rules, we accurately classified all the observations from this dataset:

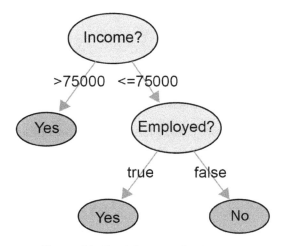

Figure 4.3: Decision tree for income

The second decision tree is less complex. At the same time, we cannot overlook the fact that the model says, *employed people with a lower income are less likely to pay back their loans*. Unfortunately, there is not enough training data available (there are only seven observations in this example), which makes it likely that we'll end up with false conclusions.

Overfitting is a frequent problem in decision trees when making a decision based on a few data points. This decision is rarely representative.

Since we can build decision trees in any possible order, it makes sense to define an efficient way of constructing a decision tree. Therefore, we will now explore a measure for ordering the features in the decision process.

ENTROPY

In information theory, entropy measures how randomly distributed the possible values of an attribute are. The higher the degree of randomness is, the higher the entropy of the attribute.

Entropy is the highest possibility of an event. If we know beforehand what the outcome will be, then the event has no randomness. So, entropy is **zero**.

We use entropy to order the splitting of nodes in the decision tree. Taking the previous example, which rule should we start with? Should it be **Income <= 75000** or **is employed**? We need to use a metric that can tell us that one specific split is better than the other. A good split can be defined by the fact it clearly split the data into two homogenous groups. One of these metrics is information gain, and it is based on entropy.

Here is the formula for calculating entropy:

$$H(\text{distribution}) = \sum_{i=0}^{n} -p_i \log_i p_i$$

Figure 4.4: Entropy formula

p_i represents the probability of one of the possible values of the target variable occurring. So, if this column has *n* different unique values, then we will have the probability for each of them *([p_1, p_2, ..., p_n])* and apply the formula.

To manually calculate the entropy of a distribution in Python, we can use the **np.log2** and **np.dot()** methods from the NumPy library. There is no function in **numpy** to automatically calculate entropy.

Have a look at the following example:

```
import numpy as np


probabilities = list(range(1,4))
minus_probabilities = [-x for x in probabilities]
log_probabilities = [x for x in map(np.log2, probabilities)]
entropy_value = np.dot(minus_probabilities, log_probabilities)
```

The probabilities are given as a NumPy array or a regular list on *line 2*: p_i.

We need to create a vector of the negated values of the distribution in *line 3*: $-p_i$.

In *line 4*, we must take the base two logarithms of each value in the distribution list: $\log_i p_i$.

Finally, we calculate the sum with the scalar product, also known as the dot product of the two vectors:

$$\sum_{i=0}^{n} -p_i \log_i p_i$$

Figure 4.5: Dot product of two vectors

NOTE

You learned about the dot product for the first time in *Chapter 2, An Introduction to Regression*. The dot product of two vectors is calculated by multiplying the i_{th} coordinate of the first vector by the i_{th} coordinate of the second vector, for each i. Once we have all the products, we sum the values:

np.dot([1, 2, 3], [4, 5, 6])

This results in 1*4 + 2*5 + 3*6 = 32.

In the next exercise, we will be calculating entropy on a small sample dataset.

EXERCISE 4.01: CALCULATING ENTROPY

In this exercise, we will calculate the entropy of the features in the dataset in *Figure 4.6*:

ID	Employed	Income	LoanType	LoanAmount	CreditWorthy
A	false	75,000	car	30,000	No
B	false	25,000	studies	15,000	No
C	true	125,000	car	30,000	Yes
D	true	75,000	car	30,000	Yes
E	true	100,000	studies	25,000	Yes
F	true	100,000	house	125,000	Yes
G	false	80,000	house	150,000	Yes

Figure 4.6: Sample dataset to formulate the rules

NOTE

The dataset file can also be found in our GitHub repository:

https://packt.live/2AQ6Uo9.

We will calculate entropy for the **Employed**, **Income**, **LoanType**, and **LoanAmount** features.

The following steps will help you complete this exercise:

1. Open a new Jupyter Notebook file.

2. Import the **numpy** package as **np**:

```
import numpy as np
```

3. Define a function called **entropy()** that receives an array of probabilities and then returns the calculated entropy value, as shown in the following code snippet:

```
def entropy(probabilities):
    minus_probabilities = [-x for x in probabilities]
    log_probabilities = [x for x in map(np.log2, \
                                    probabilities)]
    return np.dot(minus_probabilities, log_probabilities)
```

Next, we will calculate the entropy of the **Employed** column. This column contains only two possible values: **true** or **false**. The **true** value appeared four times out of seven rows, so its probability is **4/7**. Similarly, the probability of the **false** value is **3/7** as it appeared three times in this dataset.

4. Use the **entropy()** function to calculate the entropy of the **Employed** column with the probabilities **4/7** and **3/7**:

```
H_employed = entropy([4/7, 3/7])
H_employed
```

You should get the following output:

```
0.9852281360342515
```

This value is quite close to zero, which means the groups are quite homogenous.

5. Now, use the **entropy()** function to calculate the entropy of the **Income** column with its corresponding list of probabilities:

```
H_income = entropy([1/7, 2/7, 1/7, 2/7, 1/7])
H_income
```

You should get the following output:

```
2.2359263506290326
```

Compared to the **Employed** column, the entropy for **Income** is higher. This means the probabilities of this column are more spread.

6. Use the **entropy** function to calculate the entropy of the **LoanType** column with its corresponding list of probabilities:

```
H_loanType = entropy([3/7, 2/7, 2/7])
H_loanType
```

You should get the following output:

```
1.5566567074628228
```

This value is higher than 0, so the probabilities for this column are quite spread.

7. Let's use the **entropy** function to calculate the entropy of the **LoanAmount** column with its corresponding list of probabilities:

```
H_LoanAmount = entropy([1/7, 1/7, 3/7, 1/7, 1/7])
H_LoanAmount
```

You should get the following output:

```
2.128085278891394
```

The entropy for **LoanAmount** is quite high, so its values are quite random.

> **NOTE**
>
> To access the source code for this specific section, please refer to https://packt.live/37T8DVz.
>
> You can also run this example online at https://packt.live/2By7al6. You must execute the entire Notebook in order to get the desired result.

Here, you can see that the **Employed** column has the lowest entropy among the four different columns because it has the least variation in terms of values.

By completing this exercise, you've learned how to manually calculate the entropy for each column of a dataset.

INFORMATION GAIN

When we partition the data points in a dataset according to the values of an attribute, we reduce the entropy of the system.

To describe information gain, we can calculate the distribution of the labels. Initially, in *Figure 4.1*, we had five creditworthy and two not creditworthy individuals in our dataset. The entropy belonging to the initial distribution is as follows:

```
H_label = entropy([5/7, 2/7])
H_label
```

The output is as follows:

```
0.863120568566631
```

Let's see what happens if we partition the dataset based on whether the loan amount is greater than 15,000 or not:

- In group 1, we get one data point belonging to the 15,000 loan amount. This data point is not creditworthy.

- In group 2, we have five creditworthy individuals and one non-creditworthy individual.

The entropy of the labels in each group is as follows.

For group 1, we have the following:

```
H_group1 = entropy([1])
H_group1
```

The output is as follows:

```
-0.0
```

For group 2, we have the following:

```
H_group2 = entropy([5/6, 1/6])
H_group2
```

The output is as follows:

```
0.6500224216483541
```

To calculate the information gain, let's calculate the weighted average of the group entropies:

```
H_group1 * 1/7 + H_group2 * 6/7
```

The output is as follows:

```
0.5571620756985892
```

Now, to find the information gain, we need to calculate the difference between the original entropy (**H_label**) and the one we just calculated:

```
Information_gain = 0.863120568566631 - 0.5572
Information_gain
```

The output is as follows:

```
0.30592056856663097
```

By splitting the data with this rule, we gain a little bit of information.

When creating the decision tree, on each node, our job is to partition the dataset using a rule that maximizes the information gain.

We could also use Gini Impurity instead of entropy-based information gain to construct the best rules for splitting decision trees.

GINI IMPURITY

Instead of entropy, there is another widely used metric that can be used to measure the randomness of a distribution: Gini Impurity.

Gini Impurity is defined as follows:

$$\text{Gini}(\text{distribution}) = 1 - \sum_{i=0}^{n} p_i^2$$

Figure 4.7: Gini Impurity

p_i here represents the probability of one of the possible values of the target variable occurring.

Entropy may be a bit slower to calculate because of the usage of the logarithm. Gini Impurity, on the other hand, is less precise when it comes to measuring randomness.

> **NOTE**
>
> Some programmers prefer Gini Impurity because you don't have to calculate with logarithms. Computation-wise, none of the solutions are particularly complex, and so both can be used. When it comes to performance, the following study concluded that there are often just minimal differences between the two metrics: https://www.unine.ch/files/live/sites/imi/files/shared/documents/papers/Gini_index_fulltext.pdf.

With this, we have learned that we can optimize a decision tree by splitting the data based on information gain or Gini Impurity. Unfortunately, these metrics are only available for discrete values. What if the label is defined on a continuous interval such as a price range or salary range?

We have to use other metrics. You can technically understand the idea behind creating a decision tree based on a continuous label, which was about regression. One metric we can reuse in this chapter is the mean squared error. Instead of Gini Impurity or information gain, we have to minimize the mean squared error to optimize the decision tree. As this is a beginner's course, we will omit this metric.

In the next section, we will discuss the exit condition for a decision tree.

EXIT CONDITION

We can continuously split the data points according to more and more specific rules until each leaf of the decision tree has an entropy of zero. The question is whether this end state is desirable.

Often, this is not what we expect, because we risk overfitting the model. When our rules for the model are too specific and too nitpicky, and the sample size that the decision was made on is too small, we risk making a false conclusion, thus recognizing a pattern in the dataset that simply does not exist in real life.

For instance, if we spin a roulette wheel three times and we get 12, 25, and 12, this concludes that every odd spin resulting in the value 12 is not a sensible strategy. By assuming that every odd spin equals 12, we find a rule that is exclusively due to random noise.

Therefore, posing a restriction on the minimum size of the dataset that we can still split is an exit condition that works well in practice. For instance, if you stop splitting as soon as you have a dataset that's lower than 50, 100, 200, or 500 in size, you avoid drawing conclusions on random noise, and so you minimize the risk of overfitting the model.

Another popular exit condition is the maximum restriction on the depth of the tree. Once we reach a fixed tree depth, we classify the data points in the leaves.

BUILDING DECISION TREE CLASSIFIERS USING SCIKIT-LEARN

We have already learned how to load data from a **.csv** file, how to apply preprocessing to data, and how to split data into training and testing datasets. If you need to refresh yourself on this knowledge, you can go back to the previous chapters, where you can go through this process in the context of regression and classification.

Now, we will assume that a set of training features, training labels, testing features, and testing labels have been given as a return value of the **scikit-learn train-test-split** call:

```
from sklearn import model_selection
features_train, features_test, \
label_train, label_test = \
model_selection.train_test_split(features, label, test_size=0.1, \
                                 random_state=8)
```

In the preceding code snippet, we used **train_test_split** to split the dataset (features and labels) into training and testing sets. The testing set represents 10% of the observation (**test_size=0.1**). The **random_state** parameter is used to get reproducible results.

We will not focus on how we got these data points because this process is exactly the same as in the case of regression and classification.

It's time to import and use the decision tree classifier of scikit-learn:

```
from sklearn.tree import DecisionTreeClassifier
decision_tree = DecisionTreeClassifier(max_depth=6)
decision_tree.fit(features_train, label_train)
```

We set one optional parameter in **DecisionTreeClassifier**, that is, **max_depth**, to limit the depth of the decision tree.

> **NOTE**
>
> You can read the official documentation for the full list of parameters: http://scikit-learn.org/stable/modules/generated/sklearn.tree.DecisionTreeClassifier.html.

Some of the more important parameters are as follows:

- **`criterion`**: Gini stands for Gini Impurity, while entropy stands for information gain. This will define which measure will be used to assess the quality of a split at each node.

- **`max_depth`**: This is the parameter that defines the maximum depth of the tree.

- **`min_samples_split`**: This is the minimum number of samples needed to split an internal node.

You can also experiment with all the other parameters that were enumerated in the documentation. We will omit them in this section.

Once the model has been built, we can use the decision tree classifier to predict data:

```
decision_tree.predict(features_test)
```

You will build a decision tree classifier in the activity at the end of this section.

PERFORMANCE METRICS FOR CLASSIFIERS

After splitting the training and testing data, the decision tree model has a **score** method to evaluate how well testing data is classified by the model (also known as the accuracy score). We learned how to use the **score** method in the previous two chapters:

```
decision_tree.score(features_test, label_test)
```

The return value of the **score** method is a number that's less than or equal to 1. The closer we get to 1, the better our model is.

Now, we will learn about another way to evaluate the model.

> **NOTE**
>
> Feel free to use this method on the models you constructed in the previous chapter as well.

Suppose we have one test feature and one test label:

```
predicted_label = decision_tree.predict(features_test)
```

Let's use the previous creditworthy example and assume we trained a decision tree and now have its predictions:

ID	Employed	Income	LoanType	LoanAmount	CreditWorthy	Prediction
A	false	75,000	car	30,000	No	Yes
B	false	25,000	studies	15,000	No	No
C	true	125,000	car	30,000	Yes	Yes
D	true	75,000	car	30,000	Yes	No
E	true	100,000	studies	25,000	Yes	No
F	true	100,000	house	125,000	Yes	Yes
G	false	80,000	house	150,000	Yes	Yes

Figure 4.8: Sample dataset to formulate the rules

Our model, in general, made good predictions but had few errors. It incorrectly predicted the results for IDs **A**, **D**, and **E**. Its accuracy score will be 4 / 7 = 0.57.

We will use the following definitions to define some metrics that will help you evaluate how good your classifier is:

- **True positive (or TP)**: All the observations where the true label (the `Creditworthy` column, in our example) and the corresponding predictions both have the value **Yes**. In our example, IDs **C**, **F**, and **G** will fall under this category.

- **True negative (or TN)**: All the observations where the true label and the corresponding predictions both have the value **No**. Only ID **B** will be classified as true negative.

- **False positive (or FP)**: All the observations where the prediction is **Yes** but the true label is actually **No**. This will be the case for ID **A**.

- **False negative (or FN)**: All the observations where the prediction is **No** but the true label is actually **Yes**, such as for IDs **D** and **E**.

Using the preceding four definitions, we can define four metrics that describe how well our model predicts the target variable. The **# (X)** symbol denotes the number of values in **X**. Using technical terms, **# (X)** denotes the cardinality of **X**:

Definition (Accuracy): *#(True Positives) + #(True Negatives) / #(Dataset)*

Accuracy is a metric that's used for determining how many times the classifier gives us the correct answer. This is the first metric we used to evaluate the score of a classifier.

In our previous example (*Figure 4.8*), the accuracy score will be TP + TN / total = (3 + 1) / 7 = 4/7.

We can use the function provided by scikit-learn to calculate the accuracy of a model:

```
from sklearn.metrics import accuracy_score

accuracy_score(label_test, predicted_label)
```

Definition (Precision): *#TruePositives / (#TruePositives + #FalsePositives)*

Precision centers around values that our classifier found to be positive. Some of these results are true positive, while others are false positive. High precision means that the number of false positive results is very low compared to the true positive results. This means that a precise classifier rarely makes a mistake when finding a positive result.

Definition (Recall): *#True Positives / (#True Positives + #False Negatives)*

Recall centers around values that are positive among the test data. Some of these results are found by the classifier. These are the true positive values. Those positive values that are not found by the classifier are false negatives. A classifier with a high recall value finds most of the positive values.

Using our previous example (*Figure 4.8*), we will get the following measures:

- Precision = TP / (TP + FP) = 4 / (4 + 1) = 4/6 = 0.8

- Recall = TP / (TP + FN) = 4 / (4 + 2) = 4/6 = 0.6667

With these two measures, we can easily see where our model is performing better or worse. In this example, we know it tends to misclassify false negative cases. These measures are more granular than the accuracy score, which only gives you an overall score.

The F_1 score is a metric that combines precision and recall scores. Its value ranges between 0 and 1. If the F_1 score equals 1, it means the model is perfectly predicting the right outcomes. On the other hand, an F_1 score of 0 means the model cannot predict the target variable accurately. The advantage of the F_1 score is that it considers both false positives and false negatives.

The formula for calculating the F_1 score is as follows:

$$F_1 = 2 * \frac{precision * recall}{precision + recall}$$

Figure 4.9: Formula to calculate the F_1 score

As a final note, the scikit-learn package also provides a handy function that can show all these measures in one go: **classification_report()**. A classification report is useful to check the quality of our predictions:

```
from sklearn.metrics import classification_report
print(classification_report(label_test, predicted_label))
```

In the next exercise, we will be practicing how to calculate these scores manually.

EXERCISE 4.02: PRECISION, RECALL, AND F1 SCORE CALCULATION

In this exercise, we will calculate the precision, recall value, and the F_1 score of two different classifiers on a simulated dataset.

The following steps will help you complete this exercise:

1. Open a new Jupyter Notebook file.

2. Import the **numpy** package as **np** using the following code:

    ```
    import numpy as np
    ```

3. Create a **numpy** array called **real_labels** that contains the values [**True, True, False, True, True**]. This list will represent the true values of the target variable for our simulated dataset. Print its content:

    ```
    real_labels = np.array([True, True, False, True, True])
    real_labels
    ```

The expected output will be as follows:

```
array([ True, True, False, True, True])
```

4. Create a **numpy** array called **model_1_preds** that contains the values **[True, False, False, False, False]**. This list will represent the predicted values of the first classifier. Print its content:

```
model_1_preds = np.array([True, False, False, False, False])
model_1_preds
```

The expected output will be as follows:

```
array([ True, False, False, False, False])
```

5. Create another **numpy** array called **model_2_preds** that contains the values **[True, True, True, True, True]**. This list will represent the predicted values of the first classifier. Print its content:

```
model_2_preds = np.array([True, True, True, True, True])
model_2_preds
```

The expected output will be as follows:

```
array([ True,   True,   True,   True,   True])
```

6. Create a variable called **model_1_tp_cond** that will find the true positives for the first model:

```
model_1_tp_cond = (real_labels == True) \
                  & (model_1_preds == True)
model_1_tp_cond
```

The expected output will be as follows:

```
array([ True, False, False, False, False])
```

7. Create a variable called **model_1_tp** that will get the number of true positives for the first model by summing **model_1_tp_cond**:

```
model_1_tp = model_1_tp_cond.sum()
model_1_tp
```

The expected output will be as follows:

```
1
```

There is only **1** true positive case for the first model.

8. Create a variable called **model_1_fp** that will get the number of false positives for the first model:

```
model_1_fp = ((real_labels == False) \
              & (model_1_preds == True)).sum()
model_1_fp
```

The expected output will be as follows:

```
0
```

There is no false positive for the first model.

9. Create a variable called **model_1_fn** that will get the number of false negatives for the first model:

```
model_1_fn = ((real_labels == True) \
              & (model_1_preds == False)).sum()
model_1_fn
```

The expected output will be as follows:

```
3
```

The first classifier presents **3** false negative cases.

10. Create a variable called **model_1_precision** that will calculate the precision for the first model:

```
model_1_precision = model_1_tp / (model_1_tp + model_1_fp)
model_1_precision
```

The expected output will be as follows:

```
1.0
```

The first classifier has a precision score of **1**, so it didn't predict any false positives.

11. Create a variable called **model_1_recall** that will calculate the recall for the first model:

```
model_1_recall = model_1_tp / (model_1_tp + model_1_fn)
model_1_recall
```

The expected output will be as follows:

```
0.25
```

The recall score for the first model is only **0.25**, so it is predicting quite a lot of false negatives.

12. Create a variable called **model_1_f1** that will calculate the F_1 score for the first model:

```
model_1_f1 = 2*model_1_precision * model_1_recall\
             / (model_1_precision + model_1_recall)
model_1_f1
```

The expected output will be as follows:

```
0.4
```

As expected, the F_1 score is quite low for the first model.

13. Create a variable called **model_2_tp** that will get the number of true positives for the second model:

```
model_2_tp = ((real_labels == True) \
              & (model_2_preds == True)).sum()
model_2_tp
```

The expected output will be as follows:

```
4
```

There are **4** true positive cases for the second model.

14. Create a variable called **model_2_fp** that will get the number of false positives for the second model:

```
model_2_fp = ((real_labels == False) \
              & (model_2_preds == True)).sum()
model_2_fp
```

The expected output will be as follows:

```
1
```

There is only one false positive for the second model.

15. Create a variable called **model_2_fn** that will get the number of false negatives for the second model:

```
model_2_fn = ((real_labels == True) \
              & (model_2_preds == False)).sum()
model_2_fn
```

The expected output will be as follows:

```
0
```

There is no false negative for the second classifier.

16. Create a variable called **model_2_precision** that will calculate precision for the second model:

```
model_2_precision = model_2_tp / (model_2_tp + model_2_fp)
model_2_precision
```

The expected output will be as follows:

```
0.8
```

The precision score for the second model is quite high: **0.8**. It is not making too many mistakes regarding false positives.

17. Create a variable called **model_2_recall** that will calculate recall for the second model:

```
model_2_recall = model_2_tp / (model_2_tp + model_2_fn)
model_2_recall
```

The expected output will be as follows:

```
1.0
```

In terms of recall, the second classifier did a great job and didn't misclassify observations to false negatives.

18. Create a variable called **model_2_f1** that will calculate the F_1 score for the second model:

```
model_2_f1 = 2*model_2_precision*model_2_recall \
             / (model_2_precision + model_2_recall)
model_2_f1
```

The expected output will be as follows:

```
0.888888888888889
```

The F_1 score is quite high for the second model.

> **NOTE**
>
> To access the source code for this specific section, please refer to
> https://packt.live/3evqbtu.
>
> You can also run this example online at https://packt.live/2NoxLdo.
> You must execute the entire Notebook in order to get the desired result.

In this exercise, we saw how to manually calculate the precision, recall, and F_1 score for two different models. The first classifier has excellent precision but bad recall, while the second classifier has excellent recall and quite good precision.

EVALUATING THE PERFORMANCE OF CLASSIFIERS WITH SCIKIT-LEARN

The scikit-learn package provides some functions for automatically calculating the precision, recall, and F_1 score for you. You will need to import them first:

```
from sklearn.metrics import recall_score, \
precision_score, f1_score
```

To get the precision score, you will need to get the predictions from your model, as shown in the following code snippet:

```
label_predicted = decision_tree.predict(data)
precision_score(label_test, predicted_label, \
                average='weighted')
```

Calculating the **recall_score** can be done like so:

```
recall_score(label_test, label_predicted, average='weighted')
```

Calculating the **f1_score** can be done like so:

```
f1_score(label_test, predicted_label, average='weighted')
```

In the next section, we will learn how to use another tool, called the confusion matrix, to analyze the performance of a classifier.

THE CONFUSION MATRIX

Previously, we learned how to use some calculated metrics to assess the performance of a classifier. There is another very interesting tool that can help you evaluate the performance of a multi-class classification model: the confusion matrix.

A confusion matrix is a square matrix where the number of rows and columns equals the number of distinct label values (or classes). In the columns of the matrix, we place each test label value. In the rows of the matrix, we place each predicted label value.

A confusion matrix looks like this:

	Label: A	Label: B	Label: C
Prediction: A	88	3	2
Prediction: B	1	61	8
Prediction: C	16	9	28

Figure 4.10: Sample confusion matrix

In the preceding example, the first row of the confusion matrix is showing us that the model is doing the following:

- Correctly predicting class A **88** times

- Predicting class A when the true value is B **3** times

- Predicting class A when the true value is C **2** times

We can also see the scenario where the model is making a lot of mistakes when it is predicting C while the true value is A (16 times). A confusion matrix is a powerful tool to quickly and easily spot which classes your model is performing well or badly for.

The scikit-learn package provides a function to calculate and display a confusion matrix:

```
from sklearn.metrics import confusion_matrix

confusion_matrix(label_test, predicted_label)
```

In the next activity, you will be building a decision tree that will classify cars as unacceptable, acceptable, good, and very good for customers.

ACTIVITY 4.01: CAR DATA CLASSIFICATION

In this activity, you will build a reliable decision tree model that's capable of aiding a company in finding cars that clients are likely to buy. We will be assuming that the car rental agency is focusing on building a lasting relationship with its clients. Your task is to build a decision tree model that classifies cars into one of four categories: unacceptable, acceptable, good, and very good.

> **NOTE**
>
> The dataset file can also be found in our GitHub repository: https://packt.live/2V95I6h.
>
> The dataset for this activity can be accessed here: https://archive.ics.uci.edu/ml/datasets/Car+Evaluation.
>
> Citation – *Dua, D., & Graff, C.. (2017). UCI Machine Learning Repository.*

It is composed of six different features: **buying**, **maintenance**, **doors**, **persons**, **luggage_boot**, and **safety**. The target variable ranks the level of acceptability for a given car. It can take four different values: **unacc**, **acc**, **good**, and **vgood**.

The following steps will help you complete this activity:

1. Load the dataset into Python and import the necessary libaries.

2. Perform label encoding with **LabelEncoder()** from scikit-learn.

3. Extract the **label** variable using **pop()** from pandas.

4. Now, separate the training and testing data with **train_test_spit()** from scikit-learn. We will use 10% of the data as test data.

5. Build the decision tree classifier using **DecisionTreeClassifier()** and its methods, **fit()** and **predict()**.

6. Check the score of our model based on the test data with **score()**.

7. Create a deeper evaluation of the model using **classification_report()** from scikit-learn.

Expected output:

	precision	recall	f1-score	support
0	0.89	0.98	0.93	42
1	0.89	0.89	0.89	9
2	0.99	0.97	0.98	114
3	1.00	0.75	0.86	8
accuracy			0.96	173
macro avg	0.94	0.90	0.92	173
weighted avg	0.96	0.96	0.96	173

Figure 4.11: Output showing the expected classification report

> **NOTE**
>
> The solution to this activity can be found on page 353.

In the next section we will be looking at Random Forest Classifier.

RANDOM FOREST CLASSIFIER

If you think about the name random forest classifier, it can be explained as follows:

- A forest consists of multiple trees.

- These trees can be used for classification.

- Since the only tree we have used so far for classification is a decision tree, it makes sense that the random forest is a forest of decision trees.

- The random nature of the trees means that our decision trees are constructed in a randomized manner.

Therefore, we will base our decision tree construction on information gain or Gini Impurity.

Once you understand these basic concepts, you essentially know what a random forest classifier is all about. The more trees you have in the forest, the more accurate prediction is going to be. When performing prediction, each tree performs classification. We collect the results, and the class that gets the most votes wins.

Random forests can be used for regression as well as for classification. When using random forests for regression, instead of counting the most votes for a class, we take the average of the arithmetic mean (average) of the prediction results and return it. Random forests are not as ideal for regression as they are for classification, though, because the models that are used to predict values are often out of control, and often return a wide range of values. The average of these values is often not too meaningful. Managing the noise in a regression exercise is harder than in classification.

Random forests are often better than one simple decision tree because they provide redundancy. They treat outlier values better and have a lower probability of overfitting the model. Decision trees seem to behave great as long as you are using them on the data that was used when creating the model. Once you use them to predict new data, random forests lose their edge. Random forests are widely used for classification problems, whether it be customer segmentation for banks or e-commerce, classifying images, or medicine. If you own an Xbox with Kinect, your Kinect device contains a random forest classifier to detect your body.

Random forest is an ensemble algorithm. The idea behind ensemble learning is that we take an aggregated view of a decision of multiple agents that potentially have different weaknesses. Due to the aggregated vote, these weaknesses cancel out, and the majority vote likely represents the correct result.

RANDOM FOREST CLASSIFICATION USING SCIKIT-LEARN

As you may have guessed, the scikit-learn package provides an implementation of the **RandomForest** classifier with the **RandomForestClassifier** class. This class provides the exact same methods as all the scikit-learn models you have seen so far – you need to instantiate a model, then fit it with the training set with **.fit()**, and finally make predictions with **.predict()**:

```
from sklearn.ensemble import RandomForestClassifier

random_forest_classifier = RandomForestClassifier()
random_forest_classifier.fit(features_train, label_train)
labels_predicted = random_forest_classifier.predict\
                    (features_test)
```

In the next section, we will be looking at the parameterization of the random forest classifier.

THE PARAMETERIZATION OF THE RANDOM FOREST CLASSIFIER

We will be considering a subset of the possible parameters, based on what we already know, which is based on the description of constructing random forests:

- **n_estimators**: The number of trees in the random forest. The default value is 10.

- **criterion**: Use Gini or entropy to determine whether you use Gini Impurity or information gain using the entropy in each tree. This will be used to find the best split at each node.

- **max_features**: The maximum number of features considered in any tree of the forest. Possible values include an integer. You can also add some strings such as **sqrt** for the square root of the number of features.

- **max_depth**: The maximum depth of each tree.

- **min_samples_split**: The minimum number of samples in the dataset in a given node to perform a split. This may also reduce the tree's size.

- **bootstrap**: A Boolean that indicates whether to use bootstrapping on data points when constructing trees.

FEATURE IMPORTANCE

A random forest classifier gives you information on how important each feature in the data classification process is. Remember, we used a lot of randomly constructed decision trees to classify data points. We can measure how accurately these data points behave, and we can also see which features are vital when it comes to decision-making.

We can retrieve the array of feature importance scores with the following query:

```
random_forest_classifier.feature_importances_
```

In this six-feature classifier, the fourth and sixth features are clearly a lot more important than any other features. The third feature has a very low importance score.

Feature importance scores come in handy when we have a lot of features and we want to reduce the feature size to avoid the classifier getting lost in the details. When we have a lot of features, we risk overfitting the model. Therefore, reducing the number of features by dropping the least significant ones is often helpful.

CROSS-VALIDATION

Earlier, we learned how to use different metrics to assess the performance of a classifier, such as the accuracy, precision, recall, or the F_1 score on a training and testing set. The objective is to have a high score on both sets that are very close to each other. In that case, your model is performant and not prone to overfitting.

The test set is used as a proxy to evaluate whether your model can generalize well to unseen data or whether it learns patterns that are only relevant to the training set.

But in the case of having quite a few hyperparameters to tune (such as for **RandomForest**), you will have to train a lot of different models and test them on your testing set. This kind of defeats the purpose of the testing set. Think of the testing set as the final exam that will define whether you pass a subject or not. You will not be allowed to pass and repass it over and over.

One solution for avoiding using the testing set too much is creating a validation set. You will train your model on the training set and use the validation set to assess its score according to different combinations of hyperparameters. Once you find your best model, you will use the testing set to make sure it doesn't overfit too much. This is, in general, the suggested approach for any data science project.

The drawback of this approach is that you are reducing the number of observations for the training set. If you have a dataset with millions of rows, it is not a problem. But for a small dataset, this can be problematic. This is where cross-validation comes in.

The following *Figure 4.12*, shows that this is a technique where you create multiple splits of the training data. For each split, the training data is separated into folds (five, in this example) and one of the folds will be used as the validation set while the others will be used for training.

For instance, for the top split, fold 5 will be used for validation and the four other folds (1 to 4) will be used to train the model. You will follow the same process for each split. After going through each split, you will have used the entire training data and the final performance score will be the average of all the models that were trained on each split:

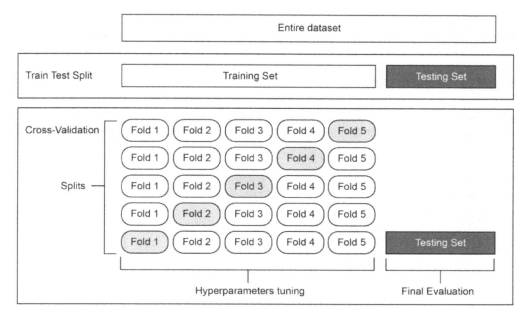

Figure 4.12: Cross-validation example

With scikit-learn, you can easily perform cross-validation, as shown in the following code snippet:

```
from sklearn.ensemble import RandomForestClassifier
random_forest_classifier = RandomForestClassifier()
from sklearn.model_selection import cross_val_score

cross_val_score(random_forest_classifier, features_train, \
                label_train, cv=5, scoring='accuracy')
```

cross_val_score takes two parameters:

- **cv**: Specifies the number of splits.

- **scoring**: Defines which performance metrics you want to use. You can find the list of possible values here: https://scikit-learn.org/stable/modules/model_evaluation.html#scoring-parameter.

In the next section, we will look at a specific variant of **RandomForest**, called **extratrees**.

EXTREMELY RANDOMIZED TREES

Extremely randomized trees increase the randomization inside random forests by randomizing the splitting rules on top of the already randomized factors in random forests.

Parameterization is like the random forest classifier. You can see the full list of parameters here: http://scikit-learn.org/stable/modules/generated/sklearn.ensemble.ExtraTreesClassifier.html.

The Python implementation is as follows:

```
from sklearn.ensemble import ExtraTreesClassifier
extra_trees_classifier = \
ExtraTreesClassifier(n_estimators=100, \
                     max_depth=6)
extra_trees_classifier.fit(features_train, label_train)
labels_predicted = extra_trees_classifier.predict(features_test)
```

In the following activity, we will be optimizing the classifier built in *Activity 4.01, Car Data Classification*.

ACTIVITY 4.02: RANDOM FOREST CLASSIFICATION FOR YOUR CAR RENTAL COMPANY

In this activity, you will optimize your classifier so that you satisfy your clients more when selecting future cars for your car fleet. We will be performing random forest and extreme random forest classification on the car dealership dataset that you worked on in the previous activity of this chapter.

The following steps will help you complete this activity:

1. Follow *Steps 1 - 4* of the previous *Activity 4.01, Car Data Classification*.

2. Create a random forest using **RandomForestClassifier**.

3. Train the models using **.fit()**.

4. Import the **confusion_matrix** function to find the quality of the **RandomForest**.

5. Print the classification report using **classification_report()**.

6. Print the feature importance with **.feature_importance_**.

7. Repeat *Steps 2 to 6* with an **extratrees** model.

Expected output:

```
array([0.08844544, 0.0702334 , 0.01440408, 0.37662014,
       0.05965896, 0.39063797])
```

> **NOTE**
>
> The solution to this activity can be found on page 357.

By completing this activity, you've learned how to fit the **RandomForest** and **extratrees** models and analyze their classification report and feature importance. Now, you can try different hyperparameters on your own and see if you can improve their results.

SUMMARY

In this chapter, we learned how to use decision trees for prediction. Using ensemble learning techniques, we created complex reinforcement learning models to predict the class of an arbitrary data point.

Decision trees proved to be very accurate on the surface, but they were prone to overfitting the model. Random forests and extremely randomized trees reduce overfitting by introducing some random elements and a voting algorithm, where the majority wins.

Beyond decision trees, random forests, and extremely randomized trees, we also learned about new methods for evaluating the utility of a model. After using the well-known accuracy score, we started using the precision, recall, and F_1 score metrics to evaluate how well our classifier works. All of these values were derived from the confusion matrix.

In the next chapter, we will describe the clustering problem and compare and contrast two clustering algorithms.

5

ARTIFICIAL INTELLIGENCE: CLUSTERING

OVERVIEW

This chapter will introduce you to the fundamentals of clustering, an unsupervised learning approach in contrast with the supervised learning approaches seen in the previous chapters. You will be implementing different types of clustering, including flat clustering with the k-means algorithm and hierarchical clustering with the mean shift algorithm and the agglomerative hierarchical model. You will also learn how to evaluate the performance of your clustering model using intrinsic and extrinsic approaches. By the end of this chapter, you will be able to analyze data using clustering and apply this skill to solve challenges across a variety of fields.

INTRODUCTION

In the previous chapter, you were introduced to decision trees and their applications in classification. You were also introduced to regression in *Chapter 2, An Introduction to Regression*. Both regression and classification are part of the supervised learning approach. However, in this chapter, we will be looking at the unsupervised learning approach; we will be dealing with datasets that don't have any labels (outputs). It is up to the machines to tell us what the labels will be based on a set of parameters that we define. In this chapter, we will be performing unsupervised learning by using clustering algorithms.

We will use clustering to analyze data to find certain patterns and create groups. Apart from that, clustering can be used for many purposes:

- Market segmentation detects the best stocks in the market you should be focusing on.

- Customer segmentation detects customer cohorts using their consumption patterns to recommend products better.

- In computer vision, image segmentation is performed using clustering. Using this, we can find different objects in an image.

- Clustering can be also be combined with classification to generate a compact representation of multiple features (inputs), which can then be fed to a classifier.

- Clustering can also filter data points by detecting outliers.

Regardless of whether we are applying clustering to genetics, videos, images, or social networks, if we analyze data using clustering, we may find similarities between data points that are worth treating uniformly.

For instance, consider a store manager, who is responsible for ensuring the profitability of their store. The products in the store are divided into different categories, and there are different customers who prefer different items. Each customer has their own preferences, but they have some similarities between them. You might have a customer who is interested in bio products, who tends to choose organic products, which are also of interest to a vegetarian customer. Even if they are different, they have similarities in their preferences or patterns as they both tend to buy organic vegetables. This can be treated as an example of clustering.

In *Chapter 3, An Introduction to Classification*, you learned about classification, which is a part of the supervised learning approach. In a classification problem, we use labels to train a model in order to be able to classify data points. With clustering, as we do not have labels for our features, we need to let the model figure out the clusters to which these features belong. This is usually based on the distance between each data point.

In this chapter, you will learn about the k-means algorithm, which is the most widely used algorithm for clustering, but first, we need to define what the clustering problem is.

DEFINING THE CLUSTERING PROBLEM

We shall define the clustering problem so that we will be able to find similarities between our data points. For instance, suppose we have a dataset that consists of points. Clustering helps us understand this structure by describing how these points are distributed.

Let's look at an example of data points in a two-dimensional space in *Figure 5.1*:

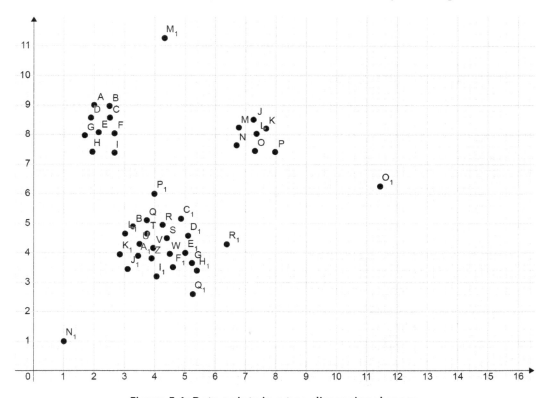

Figure 5.1: Data points in a two-dimensional space

Now, have a look, at *Figure 5.2*. It is evident that there are **three** clusters:

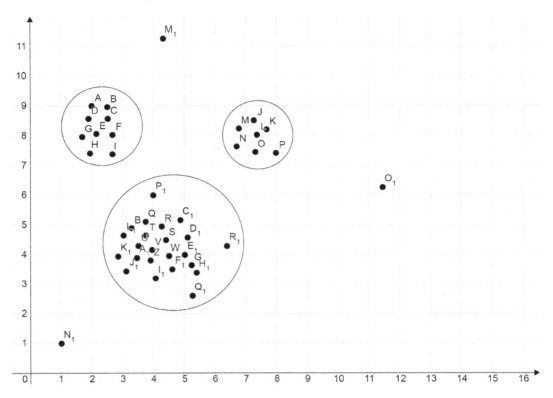

Figure 5.2: Three clusters formed using the data points in a two-dimensional space

The three clusters were easy to detect because the points are close to one another. Here, you can see that clustering determines the data points that are close to each other. You may have also noticed that the data points M_1, O_1, and N_1 do not belong to any cluster; these are the **outlier points**. The clustering algorithm you build should be prepared to treat these outlier points properly, without moving them into a cluster.

While it is easy to recognize clusters in a two-dimensional space, we normally have multidimensional data points, which is where we have more than two features. Therefore, it is important to know which data points are close to one other. Also, it is important to define the distance metrics that detect whether data points are close to each other. One well-known distance metric is Euclidean distance, which we learned about in *Chapter 1, Introduction to Artificial Intelligence*. In mathematics, we often use Euclidean distance to measure the distance between two points. Therefore, Euclidean distance is an intuitive choice when it comes to clustering algorithms so that we can determine the proximity of data points when locating clusters.

However, there is one drawback to most distance metrics, including Euclidean distance: the more we increase the dimensions, the more uniform these distances will become compared to each other. When we only have a few dimensions or features, it is easy to see which point is the closest to another one. However, when we add more features, the relevant features get embedded with all the other data and it becomes very hard to distinguish the relevant features from the others as they act as noise for our model. Therefore, getting rid of these noisy features may greatly increase the accuracy of our clustering model.

> **NOTE**
>
> Noise in a dataset can be irrelevant information or randomness that is unwanted.

In the next section, we will be looking at two different clustering approaches.

CLUSTERING APPROACHES

There are two types of clustering:

- **Flat**

- **Hierarchical**

In flat clustering, we specify the number of clusters we would like the machine to find. One example of flat clustering is the k-means algorithm, where *k* specifies the number of clusters we would like the algorithm to use.

In hierarchical clustering, however, the machine learning algorithm itself finds out the number of clusters that are needed.

Hierarchical clustering also has two approaches:

- **Agglomerative or bottom-up hierarchical clustering** treats each point as a cluster to begin with. Then, the closest clusters are grouped together. The grouping is repeated until we reach a single cluster with every data point.

- **Divisive or top-down hierarchical clustering** treats data points as if they were all in one single cluster at the start. Then the cluster is divided into smaller clusters by choosing the furthest data points. The splitting is repeated until each data point becomes its own cluster.

Figure 5.3 gives you a much more accurate description of these two clustering approaches.

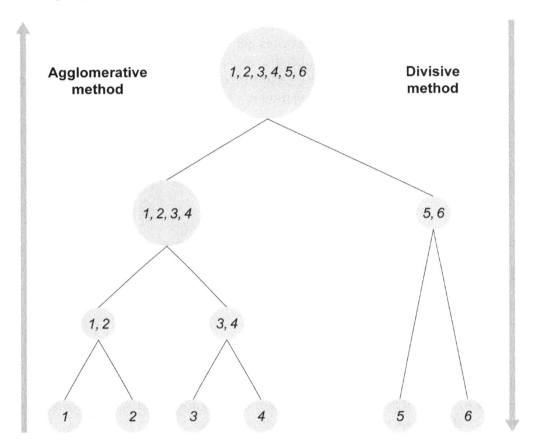

Figure 5.3: Figure showing the two approaches

Now that we are familiar with the different clustering approaches, let's take a look at the different clustering algorithms supported by scikit-learn.

CLUSTERING ALGORITHMS SUPPORTED BY SCIKIT-LEARN

In this chapter, we will learn about two clustering algorithms supported by scikit-learn:

- The k-means algorithm

- The mean shift algorithm

K-means is an example of flat clustering, where we must specify the number of clusters in advance. k-means is a general-purpose clustering algorithm that performs well if the number of clusters is not too high and the size of the clusters is uniform.

Mean shift is an example of hierarchical clustering, where the clustering algorithm determines the number of clusters. Mean shift is used when we do not know the number of clusters in advance. In contrast with k-means, mean shift supports use cases where there may be many clusters present, even if the size of the clusters greatly varies.

Scikit-learn contains many other algorithms, but we will be focusing on the k-means and mean shift algorithms in this chapter.

> **NOTE**
>
> For a complete description of clustering algorithms, including performance comparisons, visit the clustering page of scikit-learn at http://scikit-learn.org/stable/modules/clustering.html.

In the next section, we begin with the k-means algorithm.

THE K-MEANS ALGORITHM

The k-means algorithm is a flat clustering algorithm, as mentioned previously. It works as follows:

- Set the value of k.

- Choose k data points from the dataset that are the initial centers of the individual clusters.

- Calculate the distance from each data point to the chosen center points and group each point in the cluster whose initial center is the closest to the data point.

- Once all the points are in one of the k clusters, calculate the center point of each cluster. This center point does not have to be an existing data point in the dataset; it is simply an average.

- Repeat this process of assigning each data point to the cluster whose center is closest to the data point. Repetition continues until the center points no longer move.

To ensure that the k-means algorithm terminates, we need the following:

- A maximum threshold value at which the algorithm will then terminate

- A maximum number of repetitions of shifting the moving points

Due to the nature of the k-means algorithm, it will have a hard time dealing with clusters that greatly vary in size.

The k-means algorithm has many use cases that are part of our everyday lives, such as:

- **Market segmentation**: Companies gather all sorts of data on their customers. Performing k-means clustering analysis on their customers will reveal customer segments (clusters) with defined characteristics. Customers belonging to the same segment can be seen as having similar patterns or preferences.

- **Tagging of content**: Any content (videos, books, documents, movies, or photos) can be assigned tags in order to group together similar content or themes. These tags are the result of clustering.

- **Detection of fraud and criminal activities**: Fraudsters often leave clues in the form of unusual behaviors compared to other customers. For instance, in the car insurance industry, a normal customer will make a claim for a damaged car arising from an incident, whereas fraudsters will make claims for deliberate damage. Clustering can help detect whether the damage has arisen from a real accident or from a fake accident.

In the next exercise, we will be implementing the k-means algorithm in scikit-learn.

EXERCISE 5.01: IMPLEMENTING K-MEANS IN SCIKIT-LEARN

In this exercise, we will be plotting a dataset in a two-dimensional plane and performing clustering on it using the k-means algorithm.

The following steps will help you complete this exercise:

1. Open a new Jupyter Notebook file.

2. Now create an artificial dataset as a NumPy array to demonstrate the k-means algorithm. The data points are shown in the following code snippet:

```
import numpy as np
data_points = np.array([[1, 1], [1, 1.5], [2, 2], \
                        [8, 1], [8, 0], [8.5, 1], \
                        [6, 1], [1, 10], [1.5, 10], \
                        [1.5, 9.5], [10, 10], [1.5, 8.5]])
```

3. Now, plot these data points in the two-dimensional plane using **matplotlib.pyplot**, as shown in the following code snippet:

```
import matplotlib.pyplot as plot
plot.scatter(data_points.transpose()[0], \
             data_points.transpose()[1])
```

The expected output is this:

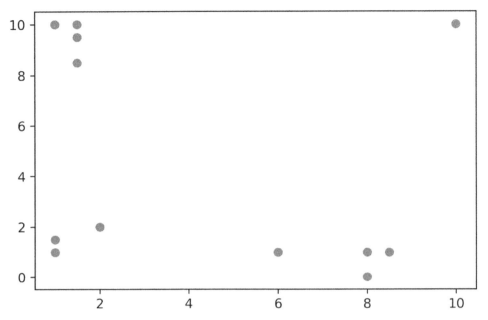

Figure 5.4: Graph showing the data points on a two-dimensional plane using matplotlib.pyplot

> **NOTE**
>
> We used the **transpose** array method to get the values of the first feature and the second feature. We could also use proper array indexing to access these columns: **dataPoints[:,0]**, which is equivalent to **dataPoints.transpose()[0]**.

Now that we have the data points, it is time to execute the k-means algorithm on them.

4. Define **k** as **3** in the k-means algorithm. We expect a cluster in the bottom-left, top-left, and bottom-right corners of the graph. Add **random_state = 8** in order to reproduce the same results:

```
from sklearn.cluster import KMeans
k_means_model = KMeans(n_clusters=3,random_state=8)
k_means_model.fit(data_points)
```

In the preceding code snippet, we have used the **KMeans** module from **sklearn.cluster**. As always with **sklearn**, we need to define a model with the parameter and then fit the model on the dataset.

The expected output is this:

```
KMeans(algorithm='auto', copy_x=True, init='k-means++',
      max_iter=300, n_clusters=3, n_init=10, n_jobs=None,
      precompute_distances='auto',
      random_state=8, tol=0.0001, verbose=0)
```

The output shows all the parameters for our k-means models, but the important ones are:

max_iter: Represents the maximum number of times the k-means algorithm will iterate through.

n_clusters: Represents the number of clusters to be formed by the k-means algorithm.

n_init: Represents the number of times the k-means algorithm will initialize a random point.

tol: Represents the threshold for checking whether the k-means algorithm can terminate.

5. Once the clustering is done, access the center point of each cluster as shown in the following code snippet:

```
centers = k_means_model.cluster_centers_
centers
```

The output of **centers** will be as follows:

```
array([[7.625     , 0.75      ],
       [3.1       , 9.6       ],
       [1.33333333, 1.5       ]])
```

This output is showing the coordinates of the center of our three clusters. If you look back at *Figure 5.4*, you will see that the center points of the clusters appear to be in the bottom-left, (**1.3, 1.5**), the top-left (**3.1, 9.6**), and the bottom-right (**7.265, 0.75**) corners of the graph. The *x* coordinate of the top-left cluster is **3.1**, most likely because it contains our outlier data point at **[10, 10]**.

6. Next, plot the clusters with different colors and their center points. To find out which data point belongs to which cluster, we must query the **labels** property of the k-means classifier, as shown in the following code snippet:

```
labels = k_means_model.labels_
labels
```

The output of **labels** will be as follows:

```
array([2, 2, 2, 0, 0, 0, 0, 1, 1, 1, 1, 1])
```

The output array shows which data point belongs to which cluster. This is all we need to plot the data.

7. Now, plot the data as shown in the following code snippet:

```
plot.scatter(centers[:,0], centers[:,1])
for i in range(len(data_points)):
    plot.plot(data_points[i][0], data_points[i][1], \
              ['k+','kx','k_'][k_means_model.labels_[i]])
plot.show()
```

In the preceding code snippets, we used the **matplotlib** library to plot the data points with the center of each coordinate. Each cluster has its marker (**x**, **+**, and **−**), and its center is represented by a filled circle.

The expected output is this:

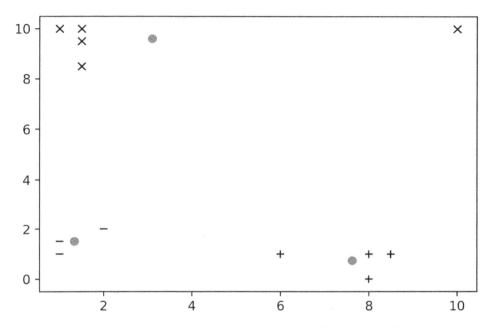

Figure 5.5: Graph showing the center points of the three clusters

Having a look at *Figure 5.5*, you can see that the center points are inside their clusters, which are represented by the **x**, **+**, and **−** marks.

8. Now, reuse the same code and choose only two clusters instead of three:

```
k_means_model = KMeans(n_clusters=2,random_state=8)
k_means_model.fit(data_points)
centers2 = k_means_model.cluster_centers_
labels2 = k_means_model.labels_
plot.scatter(centers2[:,0], centers2[:,1])
for i in range(len(data_points)):
    plot.plot(data_points[i][0], data_points[i][1], \
              ['k+','kx'][labels2[i]])
plot.show()
```

The expected output is this:

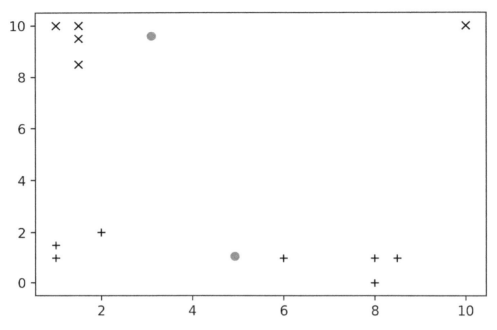

Figure 5.6: Graph showing the data points of the two clusters

This time, we only have **x** and **+** points, and we can clearly see a bottom cluster and a top cluster. Interestingly, the top cluster in the second try contains the same points as the top cluster in the first try. The bottom cluster of the second try consists of the data points joining the bottom-left and the bottom-right clusters of the first try.

9. Finally, use the k-means model for prediction as shown in the following code snippet. The output will be an array containing the cluster numbers belonging to each data point:

```
predictions = k_means_model.predict([[5,5],[0,10]])
predictions
```

The output of **predictions** is as follows:

```
array([0, 1], dtype=int32)
```

This means that our first point belongs to the first cluster (at the bottom) and the second point belongs to the second cluster (at the top).

> **NOTE**
>
> To access the source code for this specific section, please refer to https://packt.live/2CpvMDo.
>
> You can also run this example online at https://packt.live/2Nnv7F2.
> You must execute the entire Notebook in order to get the desired result.

By completing this exercise, you were able to use a simple k-means clustering model on sample data points.

THE PARAMETERIZATION OF THE K-MEANS ALGORITHM IN SCIKIT-LEARN

Like the classification and regression models in *Chapter 2*, *An Introduction to Regression*, *Chapter 3*, *An Introduction to Classification*, and *Chapter 4*, *An Introduction to Decision Trees*, the k-means algorithm can also be parameterized. The complete list of parameters can be found at http://scikit-learn.org/stable/modules/generated/sklearn.cluster.KMeans.html.

Some examples are as follows:

- **n_clusters**: The number of clusters into which the data points are separated. The default value is **8**.

- **max_iter**: The maximum number of iterations.

- **tol**: The threshold for checking whether we can terminate the k-means algorithm.

We also used two attributes to retrieve the cluster center points and the clusters themselves:

- **cluster_centers_**: This returns the coordinates of the cluster center points.

- **labels_**: This returns an array of integers representing the number of clusters the data point belongs to. Numbering starts from zero.

EXERCISE 5.02: RETRIEVING THE CENTER POINTS AND THE LABELS

In this exercise, you will be able to understand the usage of **cluster_centers_** and **labels_**.

The following steps will help you complete the exercise:

1. Open a new Jupyter Notebook file.

2. Next, create the same 12 data points from *Exercise 5.01*, *Implementing K-Means in scikit-learn*, but here, perform k-means clustering with four clusters, as shown in the following code snippet:

```
import numpy as np
import matplotlib.pyplot as plot
from sklearn.cluster import KMeans
data_points = np.array([[1, 1], [1, 1.5], [2, 2], \
                        [8, 1], [8, 0], [8.5, 1], \
                        [6, 1], [1, 10], [1.5, 10], \
                        [1.5, 9.5], [10, 10], [1.5, 8.5]])
k_means_model = KMeans(n_clusters=4,random_state=8)
k_means_model.fit(data_points)
centers = k_means_model.cluster_centers_
centers
```

The output of **centers** is as follows:

```
array([[ 7.625      ,   0.75       ],
       [ 1.375      ,   9.5        ],
       [ 1.33333333,   1.5        ],
       [10.         ,  10.         ]])
```

The output of the **cluster_centers_** property shows the *x* and *y* coordinates of the center points.

From the output, we can see the **4** centers, which are bottom right (**7.6, 0.75**), top left (**1.3, 9.5**), bottom left (**1.3, 1.5**), and top right (**10, 10**). We can also note that the fourth cluster (the top-right cluster) is only made of a single data point. This data point can be assumed to be an **outlier**.

3. Now, apply **labels_ property** on the cluster:

```
labels = k_means_model.labels_
labels
```

The output of **labels** is as follows:

```
array([2, 2, 2, 0, 0, 0, 0, 1, 1, 1, 3, 1], dtype=int32)
```

The **labels_** property is an array of length **12**, showing the cluster of each of the **12** data points it belongs to. The first cluster is associated with the number 0, the second is associated with 1, the third is associated with 2, and so on (remember that Python indexes always start from 0 and not 1).

> **NOTE**
>
> To access the source code for this specific section, please refer to https://packt.live/3dmHsDX.
>
> You can also run this example online at https://packt.live/2B0ebld.
> You must execute the entire Notebook in order to get the desired result.

By completing this exercise, you were able to retrieve the coordinates of a cluster's center. You were also able to see which label (cluster) each data point has been assigned to.

K-MEANS CLUSTERING OF SALES DATA

In the upcoming activity, we will be looking at sales data, and we will perform k-means clustering on that sales data.

ACTIVITY 5.01: CLUSTERING SALES DATA USING K-MEANS

In this activity, you will work on the Sales Transaction Dataset Weekly dataset, which contains the weekly sales data of 800 products over 1 year. Our dataset won't contain any information regarding the product except sales.

Your goal will be to identify products with similar sales trends using the k-means clustering algorithm. You will have to experiment with the number of clusters in order to find the optimal number of clusters.

> **NOTE**
>
> The dataset can be found at https://archive.ics.uci.edu/ml/datasets/Sales_Transactions_Dataset_Weekly.
>
> The dataset file can also be found in our GitHub repository: https://packt.live/3hVH42v.
>
> Citation: *Tan, S., & San Lau, J. (2014). Time series clustering: A superior alternative for market basket analysis. In Proceedings of the First International Conference on Advanced Data and Information Engineering (DaEng-2013) (pp. 241–248).*

The following steps will help you complete this activity:

1. Open a new Jupyter Notebook file.

2. Load the dataset as a DataFrame and inspect the data.

3. Create a new DataFrame without the unnecessary columns using the **drop** function from pandas (that is, the first **55** columns of the dataset) and use the **inplace** parameter, which is a part of pandas.

4. Create a k-means clustering model with **8** clusters and with **random_state = 8**.

5. Retrieve the labels from the first clustering model.

6. From the first DataFrame, **df**, keep only the **W** columns and the labels as a new column.

7. Perform the required aggregation using the **groupby** function from pandas in order to obtain the yearly average sale of each cluster.

The expected output is this:

label	count_product	total_sales	yearly_average_sales
3	115	162668	1414.504348
4	113	86037	761.389381
0	109	57020	523.119266
2	87	39940	459.080460
7	88	21213	241.056818
6	89	6137	68.955056
5	87	1375	15.804598
1	123	897	7.292683

Figure 5.7: Expected output on the Sales Transaction Data using k-means

> **NOTE**
>
> The solution to this activity is available on page 363.

Now that you have seen the k-means algorithm in detail, we will move on to another type of clustering algorithm, the mean shift algorithm.

THE MEAN SHIFT ALGORITHM

Mean shift is a hierarchical clustering algorithm that assigns data points to a cluster by calculating a cluster's center and moving it towards the mode at each iteration. The mode is the area with the most data points. At the first iteration, a random point will be chosen as the cluster's center and then the algorithm will calculate the mean of all nearby data points within a certain radius. The mean will be the new cluster's center. The second iteration will then begin with the calculation of the mean of all nearby data points and setting it as the new cluster's center. At each iteration, the cluster's center will move closer to where most of the data points are. The algorithm will stop when it is not possible for a new cluster's center to contain more data points. When the algorithm stops, each data point will be assigned to a cluster.

The mean shift algorithm will also determine the number of clusters needed, in contrast with the k-means algorithm. This is advantageous as we rarely know how many clusters we are looking for.

This algorithm also has many use cases. For instance, the Xbox Kinect device detects human body parts using the mean shift algorithm. Each main body part (head, arms, legs, hands, and so on) is a cluster of data points assigned by the mean shift algorithm.

In the next exercise, we will be implementing the mean shift algorithm.

EXERCISE 5.03: IMPLEMENTING THE MEAN SHIFT ALGORITHM

In this exercise, we will implement clustering by using the mean shift algorithm.

We will use the **scipy.spatial** library in order to compute the Euclidean distance, seen in *Chapter 1, Introduction to Artificial Intelligence*. This library simplifies the calculation of distances (such as Euclidean or Manhattan) between a list of coordinates. More details about this library can be found at https://docs.scipy.org/doc/scipy/reference/spatial.distance.html#module-scipy.spatial.distance.

The following steps will help you complete the exercise:

1. Open a new Jupyter Notebook file.

2. Let's use the data points from *Exercise 5.01, Implementing K-Means in scikit-learn*:

```
import numpy as np
data_points = np.array([[1, 1], [1, 1.5], [2, 2], \
                        [8, 1], [8, 0], [8.5, 1], \
                        [6, 1], [1, 10], [1.5, 10], \
                        [1.5, 9.5], [10, 10], [1.5, 8.5]])
```

```
import matplotlib.pyplot as plot
plot.scatter(data_points.transpose()[0], \
             data_points.transpose()[1])
```

Our task now is to find point P (x, y), for which the number of data points within radius R from point P is maximized. The points are distributed as follows:

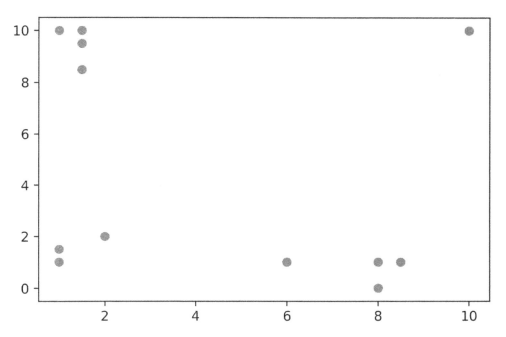

Figure 5.8: Graph showing the data points from the data_points array

3. Equate point **P1** to the first data point, **[1, 1]** of our list:

```
P1 = [1, 1]
```

4. Find the points that are within a distance of **r = 2** from this point. We will use the **scipy** library, which simplifies mathematical calculations, including spatial distance:

```
from scipy.spatial import distance
r = 2
points1 = np.array([p0 for p0 in data_points if \
                    distance.euclidean(p0, P1) <= r])
points1
```

In the preceding code snippet, we used the Euclidean distance to find all the points that fall within the **r** radius of point **P1**.

The output of **points1** will be as follows:

```
array([[1. , 1. ],
       [1. , 1.5],
       [2. , 2. ]])
```

From the output, we can see that we found three points that fall within the radius of **P1**. They are the three points at the bottom left of the graph we saw earlier, in *Figure 5.8* of this chapter.

5. Now, calculate the mean of the data points to obtain the new coordinates of **P2**:

```
P2 = [np.mean( points1.transpose()[0] ), \
      np.mean(points1.transpose()[1] )]
P2
```

In the preceding code snippet, we have calculated the mean of the array containing the three data points in order to obtain the new coordinates of **P2**.

The output of **P2** will be as follows:

```
[1.3333333333333333, 1.5]
```

6. Now that the new **P2** has been calculated, retrieve the points within the given radius again, as shown in the following code snippet:

```
points2 = np.array([p0 for p0 in data_points if \
                   distance.euclidean( p0, P2) <= r])
points2
```

The output of **points** will be as follows:

```
array([[1. , 1. ],
       [1. , 1.5],
       [2. , 2. ]])
```

These are the same three points that we found in *Step 4*, so we can stop here. Three points have been found around the mean of **[1.3333333333333333, 1.5]**. The points around this center within a radius of **2** form a cluster.

7. Since data points **[1, 1.5]** and **[2, 2]** are already in a cluster with **[1,1]**, we can directly continue with the fourth point in our list, **[8, 1]**:

```
P3 = [8, 1]
points3 = np.array( [p0 for p0 in data_points if \
                    distance.euclidean(p0, P3) <= r])
points3
```

In the preceding code snippet, we used the same code as *Step 4* but with a new **P3**.

The output of **points3** will be as follows:

```
array([[8. , 1. ],
       [8. , 0. ],
       [8.5, 1. ],
       [6. , 1. ]])
```

This time, we found four points inside the radius **r** of **P4**.

8. Now, calculate the mean, as shown in the following code snippet:

```
P4 = [np.mean(points3.transpose()[0]), \
      np.mean(points3.transpose()[1])]
P4
```

In the preceding code snippet, we calculated the mean of the array containing the four data points in order to obtain the new coordinates of **P4**, as in *Step 5*.

The output of **P4** will be as follows:

```
[7.625, 0.75]
```

This mean will not change because in the next iteration, we will find the same data points.

9. Notice that we got lucky with the selection of point **[8, 1]**. If we started with **P = [8, 0]** or **P = [8.5, 1]**, we would only find three points instead of four. Let's try with **P5 = [8, 0]**:

```
P5 = [8, 0]
points4 = np.array([p0 for p0 in data_points if \
                    distance.euclidean(p0, P5) <= r])
points4
```

In the preceding code snippet, we used the same code as in *Step 4* but with a new **P5**.

The output of **points4** will be as follows:

```
array([[8. , 1. ],
       [8. , 0. ],
       [8.5, 1. ]])
```

This time, we found three points inside the radius **r** of **P5**.

10. Now, rerun the distance calculation with the shifted mean as shown in *Step 5*:

```
P6 = [np.mean(points4.transpose()[0]), \
      np.mean(points4.transpose()[1])]
P6
```

In the preceding code snippet, we calculated the mean of the array containing the three data points in order to obtain the new coordinates of **P6**.

The output of **P6** will be as follows:

```
[8.166666666666666, 0.6666666666666666]
```

11. Now do the same again but with **P7 = [8.5, 1]**:

```
P7 = [8.5, 1]
points5 = np.array([p0 for p0 in data_points if \
                    distance.euclidean(p0, P7) <= r])
points5
```

In the preceding code snippet, we used the same code as in *Step 4* but with a new **P7**.

The output of **points5** will be as follows:

```
array([[8. , 1. ],
       [8. , 0. ],
       [8.5, 1. ]])
```

This time, we found the same three points again inside the radius **r** of **P**. This means that starting from **[8,1]**, we got a larger cluster than starting from **[8, 0]** or **[8.5, 1]**. Therefore, we must take the center point that contains the maximum number of data points.

12. Now, let's see what would happen if we started the discovery from the fourth data point, that is, **[6, 1]**:

```
P8 = [6, 1]
points6 = np.array([p0 for p0 in data_points if \
                    distance.euclidean(p0, P8) <= r])
points6
```

In the preceding code snippet, we used the same code as in *Step 4* but with a new **P8**.

The output of **points6** will be as follows:

```
array([[8., 1.],
       [6., 1.]])
```

This time, we found only two data points inside the radius **r** of **P8**. We successfully found the data point **[8, 1]**.

13. Now, shift the mean from **[6, 1]** to the calculated new mean:

```
P9 = [np.mean(points6.transpose()[0]), \
      np.mean(points6.transpose()[1]) ]
P9
```

In the preceding code snippet, we calculated the mean of the array containing the three data points in order to obtain the new coordinates of **P9**, as in *Step 5*.

The output of **P9** will be as follows:

```
[7.0, 1.0]
```

14. Check whether you have obtained more points with this new **P9**:

```
points7 = np.array([p0 for p0 in data_points if \
                    distance.euclidean(p0, P9) <= r])
points7
```

In the preceding code snippet, we used the same code as in *Step 4* but with a new **P9**.

The output of **points7** will be as follows:

```
array([[8. , 1. ],
       [8. , 0. ],
       [8.5, 1. ],
       [6. , 1. ]])
```

We successfully found all four points. Therefore, we have successfully defined a cluster of size **4**. The mean will be the same as before: `[7.625, 0.75]`.

> **NOTE**
>
> To access the source code for this specific section, please refer to https://packt.live/3drUZtE.
>
> You can also run this example online at https://packt.live/2YoSu78. You must execute the entire Notebook in order to get the desired result.

This was a simple clustering example that applied the mean shift algorithm. We only illustrated what the algorithm considers when finding clusters.

However, there is still one question, and that is what will the value of the radius be?

Note that if the radius of **2** was not set, we could simply start either with a huge radius that includes all data points and then reduce the radius, or we could start with a tiny radius, making sure that each data point is in its cluster, and then increase the radius until we get the desired result.

In the next section, we will be looking at the mean shift algorithm but using scikit-learn.

THE MEAN SHIFT ALGORITHM IN SCIKIT-LEARN

Let's use the same data points we used with the k-means algorithm:

```
import numpy as np
data_points = np.array([[1, 1], [1, 1.5], [2, 2], \
                        [8, 1], [8, 0], [8.5, 1], \
                        [6, 1], [1, 10], [1.5, 10], \
                        [1.5, 9.5], [10, 10], [1.5, 8.5]])
```

The syntax of the mean shift clustering algorithm is like the syntax for the k-means clustering algorithm:

```
from sklearn.cluster import MeanShift
mean_shift_model = MeanShift()
mean_shift_model.fit(data_points)
```

Once the clustering is done, we can access the center point of each cluster:

```
mean_shift_model.cluster_centers_
```

The expected output is this:

```
array([[ 1.375     ,  9.5       ],
       [ 8.16666667,  0.66666667],
       [ 1.33333333,  1.5       ],
       [10.        , 10.        ],
       [ 6.        ,  1.        ]])
```

The mean shift model found five clusters with the centers shown in the preceding code.

Like k-means, we can also get the labels:

```
mean_shift_model.labels_
```

The expected output is this:

```
array([2, 2, 2, 1, 1, 1, 4, 0, 0, 0, 3, 0], dtype=int64)
```

The output array shows which data point belongs to which cluster. This is all we need to plot the data:

```
import matplotlib.pyplot as plot
plot.scatter(mean_shift_model.cluster_centers_[:,0], \
             mean_shift_model.cluster_centers_[:,1])
for i in range(len(data_points)):
    plot.plot(data_points[i][0], data_points[i][1], \
             ['k+','kx','kv', 'k_', 'k1']\
             [mean_shift_model.labels_[i]])
plot.show()
```

In the preceding code snippet, we made a plot of the data points and the centers of the five clusters. Each data point belonging to the same cluster will have the same marker. The cluster centers are marked as a dot.

The expected output is this:

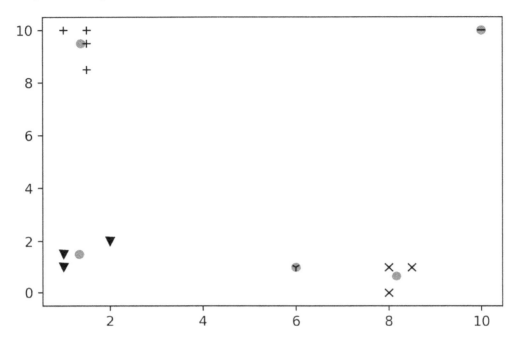

Figure 5.9: Graph showing the data points of the five clusters

We can see that three clusters contain more than a single dot (the top left, the bottom left, and the bottom right). The two single data points that are also their own cluster can be seen as outliers, as mentioned previously, as they are too far from the other clusters to be part of any of them.

Now that we have learned about the mean shift algorithm, we can have look at hierarchical clustering, and more specifically at agglomerative hierarchical clustering (the *bottom-up* approach).

HIERARCHICAL CLUSTERING

Hierarchical clustering algorithms fall into two categories:

- Agglomerative (or bottom-up) hierarchical clustering
- Divisive (or top-down) hierarchical clustering

We will only talk about agglomerative hierarchical clustering in this chapter, as it is the most widely used and most efficient of the two approaches.

Agglomerative hierarchical clustering treats each data point as a single cluster in the beginning and then successively merges (or agglomerates) the closest clusters together in pairs. In order to find the closest data clusters, agglomerative hierarchical clustering uses a heuristic such as the Euclidean or Manhattan distance to define the distance between data points. A linkage function will also be required to aggregate the distance between data points in clusters in order to define a unique value of the closeness of clusters.

Examples of linkage functions include single linkage (simple distance), average linkage (average distance), maximum linkage (maximum distance), and Ward linkage (square difference). The pairs of clusters with the smallest value of linkage will be grouped together. The grouping is repeated until we reach a single cluster containing every data point. In the end, this algorithm terminates when there is only a single cluster left.

In order to visually represent the hierarchy of clusters, a dendrogram can be used. A dendrogram is a tree where the leaves at the bottom represent data points. Each intersection between two leaves is the grouping of these two leaves. The root (top) represents a unique cluster that contains all the data points. Have a look at *Figure 5.10*, which represents a dendrogram.

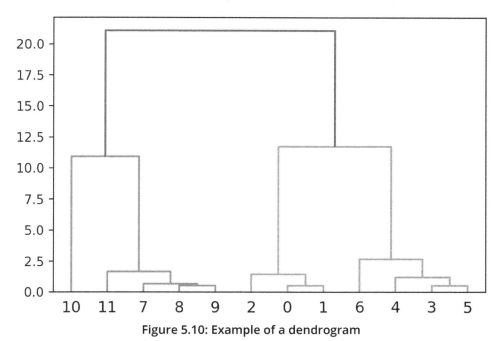

Figure 5.10: Example of a dendrogram

AGGLOMERATIVE HIERARCHICAL CLUSTERING IN SCIKIT-LEARN

Have a look at the following example, where we use the same data points as we used with the k-means algorithm:

```
import numpy as np
data_points = np.array([[1, 1], [1, 1.5], [2, 2], \
                        [8, 1], [8, 0], [8.5, 1], \
                        [6, 1], [1, 10], [1.5, 10], \
                        [1.5, 9.5], [10, 10], [1.5, 8.5]])
```

In order to plot a dendrogram, we need to first import the **scipy** library:

```
from scipy.cluster.hierarchy import dendrogram
import scipy.cluster.hierarchy as sch
```

Then we can plot a dendrogram using SciPy with the **ward** linkage function, as it is the most commonly used linkage function:

```
dendrogram = sch.dendrogram(sch.linkage(data_points, \
                            method='ward'))
```

The output of the dendrogram will be as follows:

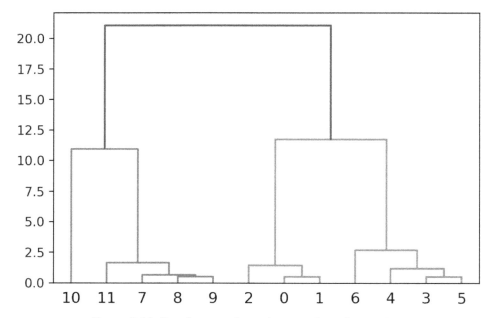

Figure 5.11: Dendrogram based on random data points

With the dendrogram, we can generally guess what will be a good number of clusters by simply drawing a horizontal line as shown in *Figure 5.12*, in the area with the highest vertical distance, and counting the number of intersections. In this case, it should be two clusters, but we will go to the next biggest area as two is too small a number.

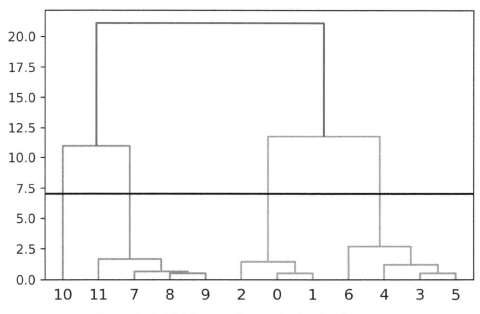

Figure 5.12: Division on clusters in the dendrogram

The *y* axis represents the measure of closeness, and the *x* axis represents the index of each data point. So our first three data points (**0**, **1**, **2**) are parts of the same cluster, then another cluster is made of the next four points (**3**, **4**, **5**, **6**), data point **10** is a cluster on its own, and the remaining data points (**7**, **8**, **9**, **11**) form the last cluster.

The syntax of the agglomerative hierarchical clustering algorithm is similar to the k-means clustering algorithm except that we need to specify the number type of **affinity** (here, we choose the Euclidean distance) and the linkage (here, we choose the **ward** linkage):

```
from sklearn.cluster import AgglomerativeClustering
agglomerative_model = AgglomerativeClustering(n_clusters=4, \
                                              affinity='euclidean', \
                                              linkage='ward')
agglomerative_model.fit(data_points)
```

The output is:

```
AgglomerativeClustering(affinity='euclidean',
                        compute_full_tree='auto',
                        connectivity=None,
                        distance_threshold=None,
                        linkage='ward', memory=None,
                        n_clusters=4, pooling_func='deprecated')
```

Similar to k-means, we can also get the labels as shown in the following code snippet:

```
agglomerative_model.labels_
```

The expected output is this:

```
array([2, 2, 2, 0, 0, 0, 0, 1, 1, 1, 3, 1], dtype=int64)
```

The output array shows which data point belongs to which cluster. This is all we need to plot the data:

```
import matplotlib.pyplot as plot
for i in range(len(data_points)):
    plot.plot(data_points[i][0], data_points[i][1], \
              ['k+','kx','kv', 'k_'][agglomerative_model.labels_[i]])
plot.show()
```

In the preceding code snippet, we made a plot of the data points and the four clusters' centers. Each data point belonging to the same cluster will have the same marker.

The expected output is this:

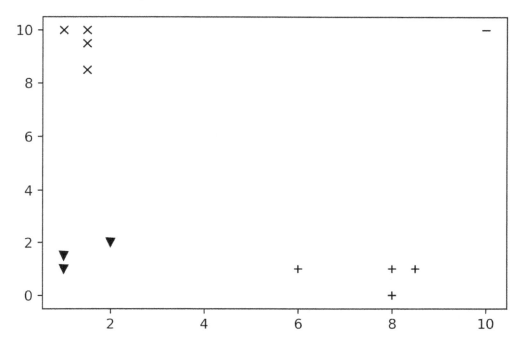

Figure 5.13: Graph showing the data points of the four clusters

We can see that, in contrast with the result from the mean shift method, agglomerative clustering was able to properly group the data point at (**6**, **1**) with the bottom-right cluster instead of having his own cluster. In situations like this one, where we have a very small amount of data, agglomerative hierarchical clustering and mean shift will work better than k-means. However, they have very expensive computational time requirements, which will make them struggle on very large datasets. However, k-means is very fast and is a better choice for very large datasets.

Now that we have learned about a few different clustering algorithms, we need to start evaluating these models and comparing them in order to choose the best model for clustering.

CLUSTERING PERFORMANCE EVALUATION

Unlike supervised learning, where we always have the labels to evaluate our predictions with, unsupervised learning is a bit more complex as we do not usually have labels. In order to evaluate a clustering model, two approaches can be taken depending on whether the label data is available or not:

- The first approach is the extrinsic method, which requires the existence of label data. This means that in absence of label data, human intervention is required in order to label the data or at least a subset of it.

- The other approach is the intrinsic approach. In general, the extrinsic approach tries to assign a score to clustering, given the label data, whereas the intrinsic approach evaluates clustering by examining how well the clusters are separated and how compact they are.

> **NOTE**
>
> We will skip the mathematical explanations as they are quite complicated.
>
> You can find more mathematical details on the sklearn website at this URL: https://scikit-learn.org/stable/modules/clustering.html#clustering-performance-evaluation

We will begin with the extrinsic approach (as it is the most widely used method) and define the following scores using sklearn on our k-means example:

- The adjusted Rand index

- The adjusted mutual information

- The homogeneity

- The completeness

- The V-Measure

- The Fowlkes-Mallows score

- The contingency matrix

Let's have a look at an example in which we first need to import the **metrics** module from **sklearn.cluster**:

```
from sklearn import metrics
```

We will be reusing the code from our k-means example in *Exercise 5.01, Implementing K-Means in scikit-learn*:

```
import numpy as np
import matplotlib.pyplot as plot
from sklearn.cluster import KMeans
data_points = np.array([[1, 1], [1, 1.5], [2, 2], \
                        [8, 1], [8, 0], [8.5, 1], \
                        [6, 1], [1, 10], [1.5, 10], \
                        [1.5, 9.5], [10, 10], [1.5, 8.5]])
k_means_model = KMeans(n_clusters=3, random_state = 8)
k_means_model.fit(data_points)
k_means_model.labels_
```

The output of our predicted labels using **k_means_model.labels_** was:

```
array([2, 2, 2, 0, 0, 0, 0, 1, 1, 1, 1, 1])
```

Finally, define the true labels of this dataset, as shown in the following code snippet:

```
data_labels = np.array([0, 0, 0, 2, 2, 2, 2, 1, 1, 1, 3, 1])
```

THE ADJUSTED RAND INDEX

The adjusted Rand index is a function that measures the similarity between the cluster predictions and the labels while ignoring permutations. The adjusted Rand index works quite well when the labels are large equal-sized clusters.

The adjusted Rand index has a range between **[-1.1]**, where negative values are not desirable. A negative score means that our model is performing worse than if we were to randomly assign labels. If we were to randomly assign them, our score would be close to 0. However, the closer we are to 1, the better our clustering model is at predicting the right label.

With **sklearn**, we can easily compute the adjusted Rand index by using this code:

```
metrics.adjusted_rand_score(data_labels, k_means_model.labels_)
```

The expected output is this:

```
0.8422939068100358
```

In this case, the adjusted Rand index indicates that our k-means model is not far from our true labels.

THE ADJUSTED MUTUAL INFORMATION

The adjusted mutual information is a function that measures the entropy between the cluster predictions and the labels while ignoring permutations.

The adjusted mutual information has no defined range, but negative values are considered bad. The closer we are to 1, the better our clustering model is at predicting the right label.

With **sklearn**, we can easily compute it by using this code:

```
metrics.adjusted_mutual_info_score(data_labels, \
                                    k_means_model.labels_)
```

The expected output is this:

```
0.8769185235006342
```

In this case, the adjusted mutual information indicates that our k-means model is quite good and not far from our true labels.

THE V-MEASURE, HOMOGENEITY, AND COMPLETENESS

The V-Measure is defined as the harmonic mean of homogeneity and completeness. The harmonic mean is a type of average (other types are the arithmetic mean and the geometric mean) using reciprocals (a reciprocal is the inverse of a number. For example the reciprocal of 2 is $\frac{1}{2}$, and the reciprocal of 3 is $\frac{1}{3}$).

The formula of the harmonic mean is as follows:

$$\frac{n}{\sum \frac{1}{x_i}}$$

Figure 5.14: The harmonic mean formula

n is the number of values and x_i is the value of each point.

In order to calculate the V-Measure, we first need to define homogeneity and completeness.

Perfect homogeneity refers to a situation where each cluster has data points belonging to the same label. The homogeneity score will reflect how well each of our clusters is grouping data from the same label.

Perfect completeness refers to the situation where all data points belonging to the same label are clustered into the same cluster. The homogeneity score will reflect how well, for each of our labels, its data points are all grouped inside the same cluster.

Hence, the formula of V-Measure is as follows:

$$v = \frac{(1+\beta) \times homogeneity \times completeness}{(\beta \times homogeneity \times completeness)}$$

Figure 5.15: The V-Measure formula

β has a default value of **1**, but it can be changed to further emphasize either homogeneity or completeness.

These three scores have a range between **[0,1]**, with **0** being the worst possible score and **1** being the perfect score.

With **sklearn**, we can easily compute these three scores by using this code:

```
metrics.homogeneity_score(data_labels, k_means_model.labels_)
metrics.completeness_score(data_labels, k_means_model.labels_)
metrics.v_measure_score(data_labels, k_means_model.labels_, \
                        beta=1)
```

The output of **homogeneity_score** is as follows:

```
0.8378758055108827
```

In this case, the homogeneity score indicates that our k-means model has clusters containing different labels.

The output of **completeness_score** is as follows:

```
1.0
```

In this case, the completeness score indicates that our k-means model has successfully put every data point of each label inside the same cluster.

The output of **v_measure_score** is as follows:

```
0.9117871871412709
```

In this case, the V-Measure indicates that our k-means model, while not being perfect, has a good score in general.

THE FOWLKES-MALLOWS SCORE

The Fowlkes-Mallows score is a metric measuring the similarity within a label cluster and the prediction of the cluster, and this is defined as the geometric mean of the precision and recall (you learned about this in *Chapter 4, An Introduction to Decision Trees*).

The formula of the Fowlkes-Mallows score is as follows:

$$FMI = \frac{TP}{\sqrt{(TP + FP)(TP + FN)}}$$

Figure 5.16: The Fowlkes-Mallows formula

Let's break this down:

- True positive (or *TP*): Are all the observations where the predictions are in the same cluster as the label cluster

- False positive (or *FP*): Are all the observations where the predictions are in the same cluster but not the same as the label cluster

- False negative (or *FN*): Are all the observations where the predictions are not in the same cluster but are in the same label cluster

The Fowlkes-Mallows score has a range between [**0, 1**], with **0** being the worst possible score and **1** being the perfect score.

With **sklearn**, we can easily compute it by using this code:

```
metrics.fowlkes_mallows_score(data_labels, k_means_model.labels_)
```

The expected output is this:

```
0.8885233166386386
```

In this case, the Fowlkes-Mallows score indicates that our k-means model is quite good and not far from our true labels.

THE CONTINGENCY MATRIX

The contingency matrix is not a score, but it reports the intersection cardinality for every true/predicted cluster pair and the required label data. It is very similar to the *Confusion Matrix* seen in *Chapter 4, An Introduction to Decision Trees*. The matrix must be the same for the label and cluster name, so we need to be careful to give our cluster the same name as our label, which was not the case with the previously seen scores.

We will modify our labels from this:

```
data_labels = np.array([0, 0, 0, 2, 2, 2, 2, 1, 1, 1, 3, 1])
```

To this:

```
data_labels = np.array([2, 2, 2, 1, 1, 1, 1, 0, 0, 0, 3, 0])
```

Then, with **sklearn**, we can easily compute the contingency matrix by using this code:

```
from sklearn.metrics.cluster import contingency_matrix
contingency_matrix(k_means_model.labels_,data_labels)
```

The output of **contingency_matrix** is as follows:

```
array([[0, 4, 0, 0],
       [4, 0, 0, 1],
       [0, 0, 3, 0]])
```

The first row of the **contingency_matrix** output indicates that there are **4** data points whose true cluster is the first cluster (**0**). The second row indicates that there are also four data points whose true cluster is the second cluster (**1**); however, an extra **1** was incorrectly predicted in this cluster, but it belongs to the fourth cluster (**3**). The third row indicates that there are three data points whose true cluster is the third cluster (**2**).

We will now look at the intrinsic approach, which is required when we do not have the label. We will define the following scores using sklearn on our k-means example:

- The Silhouette Coefficient

- The Calinski-Harabasz index

- The Davies-Bouldin index

THE SILHOUETTE COEFFICIENT

The Silhouette Coefficient is an example of an intrinsic evaluation. It measures the similarity between a data point and its cluster when compared to other clusters.

It comprises two scores:

- **a**: The average distance between a data point and all other data points in the same cluster.

- **b**: The average distance between a data point and all the data points in the nearest cluster.

The Silhouette Coefficient formula is:

$$S = \frac{b - a}{max(a, b)}$$

Figure 5.17: The Silhouette Coefficient formula

The Silhouette Coefficient has a range between [**-1,1**], with **-1** meaning an incorrect clustering. A score close to zero indicates that our clusters are overlapping. A score close to **1** indicates that all the data points are assigned to the appropriate clusters.

Then, with **sklearn**, we can easily compute the silhouette coefficient by using this code:

```
metrics.silhouette_score(data_points, k_means_model.labels_)
```

The output of **silhouette_score** is as follows:

```
0.6753568188872228
```

In this case, the Silhouette Coefficient indicates that our k-means model has some overlapping clusters, and some improvements can be made by separating some of the data points from one of the clusters.

THE CALINSKI-HARABASZ INDEX

The Calinski-Harabasz index measures how the data points inside each cluster are spread. It is defined as the ratio of the variance between clusters and the variance inside each cluster. The Calinski-Harabasz index doesn't have a range and starts from **0**. The higher the score is, the denser our clusters are. A dense cluster is an indication of a well-defined cluster.

With **sklearn**, we can easily compute it by using this code:

```
metrics.calinski_harabasz_score(data_points, k_means_model.labels_)
```

The output of **calinski_harabasz_score** is as follows:

```
19.52509172315154
```

In this case, the Calinski-Harabasz index indicates that our k-means model clusters are quite spread out and suggests that we might have overlapping clusters.

THE DAVIES-BOULDIN INDEX

The Davies-Bouldin index measures the average similarity between clusters. The similarity is a ratio of the distance between a cluster and its closest cluster and the average distance between each data point of a cluster and it's cluster's center. The Davies-Bouldin index doesn't have a range and starts from **0**. The closer the score is to **0** the better; it means the clusters are well separated, which is an indication of a good cluster.

With **sklearn**, we can easily compute the Davis-Bouldin index by using this code:

```
metrics.davies_bouldin_score(data_points, k_means_model.labels_)
```

The output of **davies_bouldin_score** is as follows:

```
0.404206621415983
```

In this case, the Calinski-Harabasz score indicates that our k-means model has some overlapping clusters and an improvement could be made by better separating some of the data points in one of the clusters.

ACTIVITY 5.02: CLUSTERING RED WINE DATA USING THE MEAN SHIFT ALGORITHM AND AGGLOMERATIVE HIERARCHICAL CLUSTERING

In this activity, you will work on the Wine Quality dataset and, more specifically, on red wine data. This dataset contains data on the quality of 1,599 red wines and the results of their chemical tests.

Your goal will be to build two clustering models (using the mean shift algorithm and agglomerative hierarchical clustering) in order to identify whether wines of similar quality also have similar physicochemical properties. You will also have to evaluate and compare the two clustering models using extrinsic and intrinsic approaches.

> **NOTE**
>
> The dataset can be found at the following URL: https://archive.ics.uci.edu/ml/datasets/Wine+Quality.
>
> The dataset file can be found on our GitHub repository at https://packt.live/2YYsxuu.
>
> Citation: *P. Cortez, A. Cerdeira, F. Almeida, T. Matos and J. Reis. Modeling wine preferences by data mining from physicochemical properties. In Decision Support Systems, Elsevier, 47(4):547-553, 2009*.

The following steps will help you complete the activity:

1. Open a new Jupyter Notebook file.

2. Load the dataset as a DataFrame with **sep = ";"** and inspect the data.

3. Create a mean shift clustering model, then retrieve the model's predicted labels and the number of clusters created.

4. Create an agglomerative hierarchical clustering model after creating a dendrogram and selecting the optimal number of clusters.

5. Retrieve the labels from the first clustering model.

6. Compute the following extrinsic approach scores for both models:

    The adjusted Rand index

    The adjusted mutual information

    The V-Measure

    The Fowlkes-Mallows score

7. Compute the following intrinsic approach scores for both models:

    The Silhouette Coefficient

    The Calinski-Harabasz index

    The Davies-Bouldin index

The expected output is this:

The values of each score for the mean shift clustering model will be as follows:

- The adjusted Rand index: **0.0006771608724007207**

- The adjusted mutual information: **0.004837187596124968**

- The V-Measure: **0.021907254751144124**

- The Fowlkes-Mallows score: **0.5721233634622408**

- The Silhouette Coefficient: **0.32769323700400077**

- The Calinski-Harabasz index: **44.62091774102674**

- The Davies-Bouldin index: **0.8106334674570222**

The values of each score for the agglomerative hierarchical clustering will be as follows:

- The adjusted Rand index: **0.05358047852603172**

- The adjusted mutual information: **0.05993098663692826**

- The V-Measure: **0.07549735446050691**

- The Fowlkes-Mallows score: `0.3300681478007641`

- The Silhouette Coefficient: `0.1591882574407987`

- The Calinski-Harabasz index: `223.5171774491095`

- The Davies-Bouldin index: `1.4975443816135114`

> **NOTE**
> The solution to this activity is available on page 368.

By completing this activity, you performed mean shift and agglomerative hierarchical clustering on multiple columns for many products. You also learned how to evaluate a clustering model with an extrinsic and intrinsic approach. Finally, you used the results of your models and their evaluation to find an answer to a real-world problem.

SUMMARY

In this chapter, we learned the basics of how clustering works. Clustering is a form of unsupervised learning where the features are given, but not the labels. It is the goal of the clustering algorithms to find the labels based on the similarity of the data points.

We also learned that there are two types of clustering, flat and hierarchical, with the first type requiring the number of clusters to find, whereas the second type finds the optimal number of clusters itself.

The k-means algorithm is an example of flat clustering, whereas mean shift and agglomerative hierarchical clustering are examples of a hierarchical clustering algorithm.

We also learned about the numerous scores to evaluate the performance of a clustering model, with the labels in the extrinsic approach or without the labels in the intrinsic approach.

In *Chapter 6*, *Neural Networks and Deep Learning*, you will be introduced to a field that has become popular in this decade due to the explosion of computation power and cheap, scalable online server capacity. This field is the science of neural networks and deep learning.

6

NEURAL NETWORKS AND DEEP LEARNING

OVERVIEW

In this chapter, you will be introduced to the final topic on neural networks and deep learning. You will be learning about TensorFlow, Convolutional Neural Networks (CNNs), and Recurrent Neural Networks (RNNs). You will use key deep learning concepts to determine creditworthiness of individuals and predict housing prices in a neighborhood. Later on, you will also implement an image classification program using the skills you learned. By the end of this chapter, you will have a firm grasp on the concepts of neural networks and deep learning.

INTRODUCTION

In the previous chapter, we learned about what clustering problems are and saw several algorithms, such as k-means, that can automatically group data points on their own. In this chapter, we will learn about neural networks and deep learning networks.

The difference between neural networks and deep learning networks is the complexity and depth of the networks. Traditionally, neural networks have only one hidden layer, while deep learning networks have more than that.

Although we will use neural networks and deep learning for supervised learning, note that neural networks can also model unsupervised learning techniques. This kind of model was actually quite popular in the 1980s, but because the computation power required was limited at the time, it's only recently that this model has been widely adopted. With the democratization of Graphics Processing Units (GPUs) and cloud computing, we now have access to a tremendous amount of computation power. This is the main reason why neural networks and especially deep learning are hot topics again.

Deep learning can model more complex patterns than traditional neural networks, and so deep learning is more widely used nowadays in computer vision (in applications such as face detection and image recognition) and natural language processing (in applications such as chatbots and text generation).

ARTIFICIAL NEURONS

Artificial Neural Networks (**ANNs**), as the name implies, try to replicate how a human brain works, and more specifically how neurons work.

A neuron is a cell in the brain that communicates with other cells via electrical signals. Neurons can respond to stimuli such as sound, light, and touch. They can also trigger actions such as muscle contractions. On average, a human brain contains 10 to 20 billion neurons. That's a pretty huge network, right? This is the reason why humans can achieve so many amazing things. This is also why researchers have tried to emulate how the brain operates and in doing so created ANNs.

ANNs are composed of multiple artificial neurons that connect to each other and form a network. An artificial neuron is simply a processing unit that performs mathematical operations on some inputs ($x1$, $x2$, ..., xn) and returns the final results (y) to the next unit, as shown here:

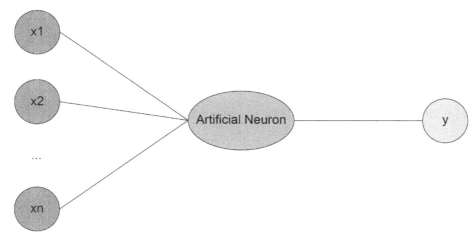

Figure 6.1: Representation of an artificial neuron

We will see how an artificial neuron works more in detail in the coming sections.

NEURONS IN TENSORFLOW

TensorFlow is currently the most popular neural network and deep learning framework. It was created and is maintained by Google. TensorFlow is used for voice recognition and voice search, and it is also the brain behind translate.google.com. Later in this chapter, we will use TensorFlow to recognize written characters.

The TensorFlow API is available in many languages, including Python, JavaScript, Java, and C. TensorFlow works with **tensors**. You can think of a tensor as a container composed of a matrix (usually with high dimensions) and additional information related to the operations it will perform (such as weights and biases, which you will be looking at later in this chapter). A tensor with no dimensions (with no rank) is a scalar. A tensor of rank 1 is a vector, rank 2 tensors are matrices, and a rank 3 tensor is a three-dimensional matrix. The rank indicates the dimensions of a tensor. In this chapter, we will be looking at tensors of ranks 2 and 3.

> **NOTE**
>
> Mathematicians use the terms matrix and dimension, whereas deep learning programmers use tensor and rank instead.

TensorFlow also comes with mathematical functions to transform tensors, such as the following:

- **Arithmetic operations**: `add` and `multiply`

- **Exponential operations**: `exp` and `log`

- **Relational operations**: `greater`, `less`, and `equal`

- **Array operations**: `concat`, `slice`, and `split`

- **Matrix operations**: `matrix_inverse`, `matrix_determinant`, and `matmul`

- **Non-linear operations**: `sigmoid`, `relu`, and `softmax`

We will go through them in more detail later in this chapter.

In the next exercise, we will be using TensorFlow to compute an artificial neuron.

EXERCISE 6.01: USING BASIC OPERATIONS AND TENSORFLOW CONSTANTS

In this exercise, we will be using arithmetic operations in TensorFlow to emulate an artificial neuron by performing a matrix multiplication and addition, and applying a non-linear function, **sigmoid**.

The following steps will help you complete the exercise:

1. Open a new Jupyter Notebook file.

2. Import the **tensorflow** package as **tf**:

    ```
    import tensorflow as tf
    ```

3. Create a tensor called **W** of shape **[1,6]** (that is, with 1 row and 6 columns), using **tf.constant()**, that contains the matrix **[1.0, 2.0, 3.0, 4.0, 5.0, 6.0]**. Print its value:

    ```
    W = tf.constant([1.0, 2.0, 3.0, 4.0, 5.0, 6.0], shape=[1, 6])
    W
    ```

    The expected output is this:

    ```
    <tf.Tensor: shape=(1, 6), dtype=float32, numpy=array([[1., 2., 3., 4.,
    5., 6.]], dtype=float32)>
    ```

4. Create a tensor called **X** of shape **[6,1]** (that is, with 6 rows and 1 column), using **tf.constant()**, that contains **[7.0, 8.0, 9.0, 10.0, 11.0, 12.0]**. Print its value:

```
X = tf.constant([7.0, 8.0, 9.0, 10.0, 11.0, 12.0], \
                    shape=[6, 1])
X
```

The expected output is this:

```
<tf.Tensor: shape=(6, 1), dtype=float32, numpy=
array([[ 7.],
       [ 8.],
       [ 9.],
       [10.],
       [11.],
       [12.]], dtype=float32)>
```

5. Now, create a tensor called **b**, using **tf.constant()**, that contains **-88**. Print its value:

```
b = tf.constant(-88.0)
b
```

The expected output is this:

```
<tf.Tensor: shape=(), dtype=float32, numpy=-88.0>
```

6. Perform a matrix multiplication between **W** and **X** using **tf.matmul**, save its results in the **mult** variable, and print its value:

```
mult = tf.matmul(W, X)
mult
```

The expected output is this:

```
<tf.Tensor: shape=(1, 1), dtype=float32, numpy=array([[217.]],
dtype=float32)>
```

7. Perform a matrix addition between **mult** and **b**, save its results in a variable called **Z**, and print its value:

```
Z = mult + b
Z
```

The expected output is this:

```
<tf.Tensor: shape=(1, 1), dtype=float32, numpy=array([[129.]],
dtype=float32)>
```

8. Apply the **sigmoid** function to **Z** using **tf.math.sigmoid**, save its results in a variable called **a**, and print its value. The **sigmoid** function transforms any numerical value within the range **0** to **1** (we will learn more about this in the following sections):

```
a = tf.math.sigmoid(Z)
a
```

The expected output is this:

```
<tf.Tensor: shape=(1, 1), dtype=float32, numpy=array([[1.]],
dtype=float32)>
```

The **sigmoid** function has transformed the original value of **Z**, which was **129**, to **1**.

> **NOTE**
>
> To access the source code for this specific section, please refer to
> https://packt.live/31ekGLM.
>
> You can also run this example online at https://packt.live/3evuKnC. You must execute the entire Notebook in order to get the desired result.

In this exercise, you successfully implemented an artificial neuron using TensorFlow. This is the base of any neural network model.

In the next section, we will be looking at the architecture of neural networks.

NEURAL NETWORK ARCHITECTURE

Neural networks are the newest branch of **Artificial Intelligence** (**AI**). Neural networks are inspired by how the human brain works. They were invented in the 1940s by Warren McCulloch and Walter Pitts. The neural network was a mathematical model that was used to describe how the human brain can solve problems.

We will use ANN to refer to both the mathematical model, and the biological neural network when talking about the human brain.

The way a neural network learns is more complex compared to other classification or regression models. The neural network model has a lot of internal variables, and the relationship between the input and output variables may involve multiple internal layers. Neural networks have higher accuracy than other supervised learning algorithms.

> **NOTE**
>
> Mastering neural networks with TensorFlow is a complex process. The purpose of this section is to provide you with an introductory resource to get started.

In this chapter, the main example we are going to use is the recognition of digits from an image. We are considering this format since each image is small, and we have around 70,000 images available. The processing power required to process these images is similar to that of a regular computer.

ANNs work similarly to how the human brain works. A dendroid in a human brain is connected to a nucleus, and the nucleus is connected to an axon. In an ANN, the input is the dendroid, where the calculations occur is the nucleus, and the output is the axon.

An artificial neuron is designed to replicate how a nucleus works. It will transform an input signal by calculating a matrix multiplication followed by an activation function. If this function determines that a neuron has to fire, a signal appears in the output. This signal can be the input of other neurons in the network:

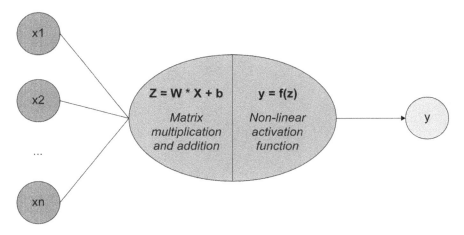

Figure 6.2: Figure showing how an ANN works

Let's understand the preceding figure further by taking the example of **n=4**. In this case, the following applies:

- **X** is the input matrix, which is composed of **x1**, **x2**, **x3**, and **x4**.

- **W**, the weight matrix, will be composed of **w1**, **w2**, **w3**, and **w4**.

- **b** is the bias.

- **f** is the activation function.

We will first calculate **z** (the left-hand side of the neuron) with matrix multiplication and bias:

```
Z = W * X + b = x1*w1 + x2*w2 + x3*w3 + x4*w4 + b
```

Then the output, **y**, will be calculated by applying a function, **f**:

```
y = f(Z) = f(x1*w1 + x2*w2 + x3*w3 + x4*w4 + b)
```

Great – this is how an artificial neuron works under the hood. It is two matrix operations, a product followed by a sum, and a function transformation.

We now move on to the next section – weights.

WEIGHTS

W (*also called the weight matrix*) refers to weights, which are parameters that are automatically learned by neural networks in order to predict accurately the output, **y**.

A single neuron is the combination of the weighted sum and the activation function and can be referred to as a hidden layer. A neural network with one hidden layer is called a **regular neural network**:

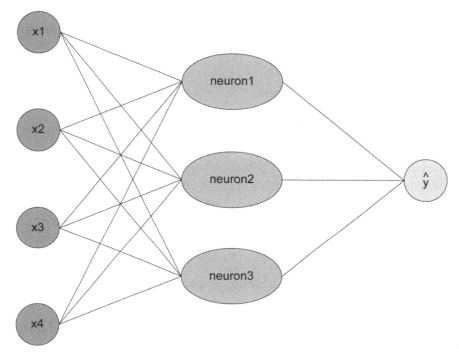

Figure 6.3: Neurons 1, 2, and 3 form the hidden layer of this sample network

When connecting inputs and outputs, we may have multiple hidden layers. A neural network with multiple layers is called a **deep neural network**.

The term deep learning comes from the presence of multiple layers. When creating an **Artificial Neural Network (ANN)**, we can specify the number of hidden layers.

BIASES

Previously, we saw that the equation for a neuron is as follows:

```
y = f(x1*w1 + x2*w2 + x3*w3 + x4*w4)
```

The problem with this equation is that there is no constant factor that depends on the inputs **x1**, **x2**, **x3**, and **x4**. The preceding equation can model any linear function that will go through the point 0: if all **w** values are equal to 0 then **y** will also equal to 0. But what about other functions that don't go through the point 0? For example, imagine that we are predicting the probability of churn for an employee by their month of tenure. Even if they haven't worked for the full month yet, the probability of churn is not zero.

To accommodate this situation, we need to introduce a new parameter called **bias**. It is a constant that is also referred to as the **intercept**. Using the churn example, the bias **b** can equal to 0.5 and therefore the churn probability for a new employer during the first month will be 50%.

Therefore, we add bias to the equation:

```
y = f(x1*w1 + x2*w2 + x3*w3 + x4*w4 + b)
y = f(x   w + b)
```

The first equation is the verbose form, describing the role of each coordinate, weight coefficient, and bias. The second equation is the vector form, where **x = (x1, x2, x3, x4)** and **w = (w1, w2, w3, w4)**. The dot operator between the vectors symbolizes the dot or scalar product of the two vectors. The two equations are equivalent. We will use the second form in practice because it is easier to define a vector of variables using TensorFlow than to define each variable one by one.

Similarly, for **w1**, **w2**, **w3**, and **w4**, the bias, **b**, is a variable, meaning that its value can change during the learning process.

With this constant factor built into each neuron, a neural network model becomes more flexible in terms of fitting a specific training dataset better.

> **NOTE**
>
> It may happen that the product **p = x1*w1 + x2*w2 + x3*w3 + x4*w4** is negative due to the presence of a few negative weights. We may still want to give the model the flexibility to execute (*or fire*) a neuron with values above a given negative number. Therefore, adding a constant bias, **b = 5**, for instance, can ensure that the neuron fires for values between −5 and 0 as well.

TensorFlow provides the **Dense ()** class to model the hidden layer of a neural network (*also called the fully connected layer*):

```
from tensorflow.keras import layers
layer1 = layers.Dense(units=128, input_shape=[200])
```

In this example, we have created a fully connected layer of **128** neurons that takes as input a tensor of shape **200**.

NOTE

You can find more information on this TensorFlow class at https://www.tensorflow.org/api_docs/python/tf/keras/layers/Dense.

The **Dense()** class is expected to have a flattened input (only one row). For instance, if your input is of shape **28** by **28**, you will have to flatten it beforehand with the **Flatten()** class in order to get a single row with 784 neurons (**28 * 28**):

```
from tensorflow.keras import layers

input_layer = layers.Flatten(input_shape=(28, 28))
layer1 = layers.Dense(units=128)
```

NOTE

You can find more information on this TensorFlow class at https://www.tensorflow.org/api_docs/python/tf/keras/layers/Flatten.

In the following sections, we will learn about how we can extend this layer of neurons with additional parameters.

USE CASES FOR ANNS

ANNs have their place among supervised learning techniques. They can model both classification and regression problems. A classifier neural network seeks a relationship between features and labels. The features are the input variables, while each class the classifier can choose as a return value is a separate output. In the case of regression, the input variables are the features, while there is one single output: the predicted value. While traditional classification and regression techniques have their use cases in AI, ANNs are generally better at finding complex relationships between inputs and outputs.

In the next section, we will be looking at activation functions and their different types.

ACTIVATION FUNCTIONS

As seen previously, a single neuron needs to perform a transformation by applying an activation function. Different activation functions can be used in neural networks. Without these functions, a neural network would simply be a linear model that could easily be described using matrix multiplication.

The activation function of a neural network provides non-linearity and therefore can model more complex patterns. Two very common activation functions are **sigmoid** and **tanh** (the hyperbolic tangent function).

SIGMOID

The formula of **sigmoid** is as follows:

$$f(x) = \sigma(x) = \frac{1}{1 + e^{-x}}$$

Figure 6.4: The sigmoid formula

The output values of a **sigmoid** function range from **0** to **1**. This activation function is usually used at the last layer of a neural network for a binary classification problem.

TANH

The formula of the hyperbolic tangent is as follows:

$$f(x) = \tanh(x) = \frac{(e^x - e^{-x})}{(e^x + e^{-x})}$$

Figure 6.5: The tanh formula

The **tanh** activation function is very similar to the **sigmoid** function and was quite popular until recently. It is usually used in the hidden layers of a neural network. Its values range between **-1** and **1**.

RELU

Another important activation function is **relu**. **ReLU** stands for **Rectified Linear Unit**. It is currently the most widely used activation function for hidden layers. Its formula is as follows:

$$f(x) = \begin{cases} 0 & \text{for } x \leq 0 \\ x & \text{for } x > 0 \end{cases}$$

Figure 6.6: The ReLU formula

There are now different variants of **relu** functions, such as **leaky ReLU** and **PReLU**.

SOFTMAX

The function shrinks the values of a list to be between **0** and **1** so that the sum of the elements of the list becomes **1**. The definition of the **softmax** function is as follows:

$$f_i(\vec{x}) = \frac{e^{x_i}}{\sum_{j=1}^{J} e^{x_j}} \text{ for } i = 1, ..., J$$

Figure 6.7: The softmax formula

The **softmax** function is usually used as the last layer of a neural network for multi-class classification problems as it can generate probabilities for each of the different output classes.

Remember, in TensorFlow, we can extend a **Dense()** layer with an activation function; we just need to set the **activation** parameter. In the following example, we will add the **relu** activation function:

```
from tensorflow.keras import layers

layer1 = layers.Dense(units=128, input_shape=[200], \
                      activation='relu')
```

Let's use these different activation functions and observe how these functions dampen the weighted inputs by solving the following exercise.

EXERCISE 6.02: ACTIVATION FUNCTIONS

In this exercise, we will be implementing the following activation functions using the **numpy** package: **sigmoid**, **tanh**, **relu**, and **softmax**.

The following steps will help you complete the exercise:

1. Open a new Jupyter Notebook file.

2. Import the **numpy** package as **np**:

```
import numpy as np
```

3. Create a **sigmoid** function, as shown in the following code snippet, that implements the sigmoid formula (shown in the previous section) using the **np.exp()** method:

```
def sigmoid(x):
    return 1 / (1 + np.exp(-x))
```

4. Calculate the result of **sigmoid** function on the value **-1**:

```
sigmoid(-1)
```

The expected output is this:

```
0.2689414213699951
```

This is the result of performing a sigmoid transformation on the value **-1**.

5. Import the **matplotlib.pyplot** package as **plt**:

```
import matplotlib.pyplot as plt
```

6. Create a **numpy** array called **x** that contains values from **-10** to **10** evenly spaced by an increment of **0.1**, using the **np.arange()** method. Print its value:

```
x = np.arange(-10, 10, 0.1)
x
```

The expected output is this:

```
array([-1.00000000e+01, -9.90000000e+00, -9.80000000e+00,
       -9.70000000e+00, -9.60000000e+00, -9.50000000e+00,
       -9.40000000e+00, -9.30000000e+00, -9.20000000e+00,
       -9.10000000e+00, -9.00000000e+00, -8.90000000e+00,
       -8.80000000e+00, -8.70000000e+00, -8.60000000e+00,
       -8.50000000e+00, -8.40000000e+00, -8.30000000e+00,
```

```
-8.20000000e+00, -8.10000000e+00, -8.00000000e+00,
-7.90000000e+00, -7.80000000e+00, -7.70000000e+00,
-7.60000000e+00, -7.50000000e+00, -7.40000000e+00,
-7.30000000e+00, -7.20000000e+00, -7.10000000e+00,
-7.00000000e+00, -6.90000000e+00,
```

Great – we generated a **numpy** array containing values between **–10** and **10**.

> **NOTE**
>
> The preceding output is truncated.

7. Plot a line chart with **x** and **sigmoid(x)** using **plt.plot()** and **plt.show()**:

```
plt.plot(x, sigmoid(x))
plt.show()
```

The expected output is this:

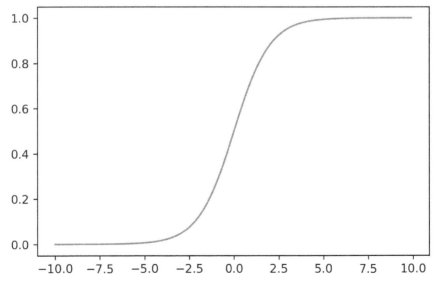

Figure 6.8: Line chart using the sigmoid function

We can see here that the output of the **sigmoid** function ranges between **0** and **1**. The slope is quite steep for values around **0**.

8. Create a **tanh()** function that implements the Tanh formula (shown in the previous section) using the **np.exp()** method:

```
def tanh(x):
    return 2 / (1 + np.exp(-2*x)) - 1
```

9. Plot a line chart with **x** and **tanh(x)** using **plt.plot()** and **plt.show()**:

```
plt.plot(x, tanh(x))
plt.show()
```

The expected output is this:

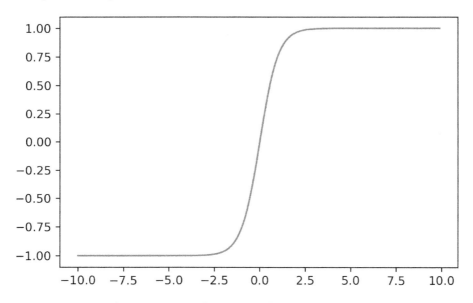

Figure 6.9: Line chart using the tanh function

The shape of the **tanh** function is very similar to **sigmoid** but its slope is steeper for values close to **0**. Remember, its range is between **-1** and **1**.

10. Create a **relu** function that implements the ReLU formula (shown in the previous section) using the **np.maximum()** method:

```
def relu(x):
    return np.maximum(0, x)
```

11. Plot a line chart with **x** and **relu(x)** using **plt.plot()** and **plt.show()**:

```
plt.plot(x, relu(x))
plt.show()
```

The expected output is this:

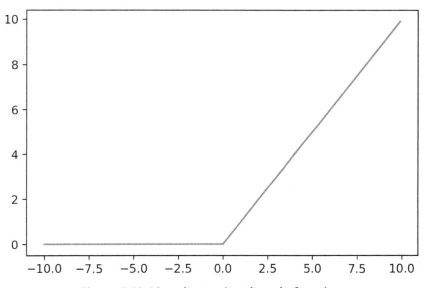

Figure 6.10: Line chart using the relu function

The ReLU function equals **0** when values are negative, and equals the identity function, **f(x)=x**, for positive values.

12. Create a **softmax** function that implements the softmax formula (shown in the previous section) using the **np.exp()** method:

```
def softmax(list):
    return np.exp(list) / np.sum(np.exp(list))
```

13. Calculate the output of **softmax** on the list of values, **[0, 1, 168, 8, 2]**:

```
result = softmax( [0, 1, 168, 8, 2])
result
```

The expected output is this:

```
array([1.09276566e-73, 2.97044505e-73, 1.00000000e+00,
       3.25748853e-70, 8.07450679e-73])
```

As expected, the item at the third position has the highest softmax probabilities as its original value was the highest.

> **NOTE**
>
> To access the source code for this specific section, please refer to https://packt.live/3fJzoOU.
>
> You can also run this example online at https://packt.live/3188pZi. You must execute the entire Notebook in order to get the desired result.

By completing this exercise, we have implemented some of the most important activation functions for neural networks.

FORWARD PROPAGATION AND THE LOSS FUNCTION

So far, we have seen how a neuron can take an input and perform some mathematical operations on it and get an output. We learned that a neural network is a combination of multiple layers of neurons.

The process of transforming the inputs of a neural network into a result is called **forward propagation** (or the forward pass). What we are asking the neural network to do is to make a prediction (the final output of the neural network) by applying multiple neurons to the input data:

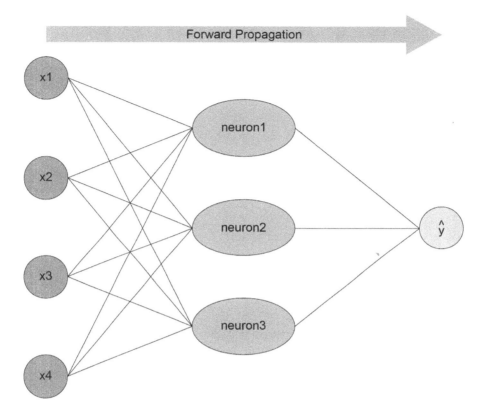

Figure 6.11: Figure showing forward propagation

The neural network relies on the weights matrices, biases, and activation function of each neuron to calculate the predicted output value, $(\hat{y})$. For now, let's assume the values of the weight matrices and biases are set in advance. The activation functions are defined when you design the architecture of the neural networks.

As for any supervised machine learning algorithm, the goal is to make accurate predictions. This implies that we need to assess how accurate the predictions are compared to the true values. For traditional machine learning algorithms, we used scoring metrics such as mean squared error, accuracy, or the F_1 score. This can also be applied to neural networks, but the only difference is that such scores are used in two different ways:

- They are used by data scientists to assess the performance of a model on training and testing sets and then tune hyperparameters if needed. This also applies to neural networks, so nothing new here.

- They are used by neural networks to automatically learn from mistakes and update weight matrices and biases. This will be explained in more detail in the next section, which is about backpropagation. So, the neural network will use a metric (also called a **loss function**) to compare its predicted values, $(\hat{y})$ to the true label, (y), and then learn how to make better predictions automatically.

The loss function is critical to a neural network learning to make good predictions. This is a hyperparameter that needs to be defined by data scientists while designing the architecture of a neural network. The choice of which loss function to use is totally arbitrary and depending on the dataset or the problem you want to solve, you will pick one or another. Luckily for us, though, there are some basic rules of thumb that work in most cases:

- If you are working on a regression problem, you can use mean squared error.

- If it is a binary classification, the loss function should be binary cross-entropy.

- If it is a multi-class classification, then categorical cross-entropy should be your go-to choice.

As a final note, the choice of loss function will also define which activation function you will have to use on the last layer of the neural network. Each loss function expects a certain type of data in order to properly assess prediction performance.

Here is the list of activation functions according to the loss function and type of project/problem:

Problem Type	Last-Layer Activation Function	Loss Function
Regression	None (or identity function)	Mean squared error
Binary classification	sigmoid	Binary cross-entropy
Multi-class classification	softmax	Categorical cross-entropy

Figure 6.12: Overview of the different activation functions and their applications

With TensorFlow, in order to build your custom architecture, you can instantiate the **Sequential()** class and add your layers of fully connected neurons as shown in the following code snippet:

```
import tensorflow as tf
from tensorflow.keras import layers

model = tf.keras.Sequential()
input_layer = layers.Flatten(input_shape=(28,28))
layer1 = layers.Dense(128, activation='relu')
```

```
model.add(input_layer)
model.add(layer1)
```

Now it is time to have a look at how a neural network improves its predictions with backpropagation.

BACKPROPAGATION

Previously, we learned how a neural network makes predictions by using weight matrices and biases (we can combine them into a single matrix) from its neurons. Using the loss function, a network determines how good or bad the predictions are. It would be great if it could use this information and update the parameters accordingly. This is exactly what backpropagation is about: optimizing a neural network's parameters.

Training a neural network involves executing forward propagation and backpropagation multiple times in order to make predictions and update the parameters from the errors. During the first pass (or propagation), we start by initializing all the weights of the neural network. Then, we apply forward propagation, followed by backpropagation, which updates the weights.

We apply this process several times and the neural network will optimize its parameters iteratively. You can decide to stop this learning process by setting the maximum number of times the neural networks will go through the entire dataset (also called epochs) or define an early stop threshold if the neural network's score is not improving anymore after few epochs.

OPTIMIZERS AND THE LEARNING RATE

In the previous section, we saw that a neural network follows an iterative process to find the best solution for any input dataset. Its learning process is an optimization process. You can use different optimization algorithms (also called **optimizers**) for a neural network. The most popular ones are **Adam**, **SGD**, and **RMSprop**.

One important parameter for the neural networks optimizer is the learning rate. This value defines how quickly the neural network will update its weights. Defining a too-low learning rate will slow down the learning process and the neural network will take a long time before finding the right parameters. On the other hand, having too-high a learning rate can make the neural network not learn a solution as it is making bigger weight changes than required. A good practice is to start with a not-too-small learning rate (such as **0.01** or **0.001**), then stop the neural network training once its score starts to plateau or get worse, and lower the learning rate (by an order of magnitude, for instance) and keep training the network.

With TensorFlow, you can instantiate an optimizer from **tf.keras.optimizers**. For instance, the following code snippet shows us how to create an **Adam** optimizer with **0.001** as the learning rate and then compile our neural network by specifying the loss function (**'sparse_categorical_crossentropy'**) and metrics to be displayed (**'accuracy'**):

```
import tensorflow as tf

optimizer = tf.keras.optimizers.Adam(0.001)
model.compile(loss='sparse_categorical_crossentropy', \
              optimizer=optimizer, metrics=['accuracy'])
```

Once the model is compiled, we can then train the neural network with the **.fit()** method like this:

```
model.fit(features_train, label_train, epochs=5)
```

Here we trained the neural network on the training set for **5** epochs. Once trained, we can use the model on the testing set and assess its performance with the **.evaluate()** method:

```
model.evaluate(features_test, label_test)
```

> **NOTE**
>
> You can find more information on this TensorFlow optimizers at https://www.tensorflow.org/api_docs/python/tf/keras/optimizers.

In the next exercise, we will be training a neural network on a dataset.

EXERCISE 6.03: CLASSIFYING CREDIT APPROVAL

In this exercise, we will be using the German credit approval dataset, and train a neural network to classify whether an individual is creditworthy or not.

> **NOTE**
>
> The dataset file can also be found in our GitHub repository:
>
> https://packt.live/2V7uiV5.

The following steps will help you complete the exercise:

1. Open a new Jupyter Notebook file.

2. Import the **loadtxt** method from **numpy**:

    ```
    from numpy import loadtxt
    ```

3. Create a variable called **file_url** containing the link to the raw dataset:

    ```
    file_url = 'https://raw.githubusercontent.com/'\
               'PacktWorkshops/'\
               'The-Applied-Artificial-Intelligence-Workshop'\
               '/master/Datasets/german_scaled.csv'
    ```

4. Load the data into a variable called **data** using **loadtxt()** and specify the **delimiter=','** parameter. Print its content:

    ```
    data = loadtxt(file_url, delimiter=',')
    data
    ```

    The expected output is this:

    ```
    array([[0.        , 0.33333333, 0.02941176, ..., 0.        , 1.        ,
            1.        ],
           [1.        , 0.        , 0.64705882, ..., 0.        , 0.        ,
            1.        ],
           [0.        , 1.        , 0.11764706, ..., 1.        , 0.        ,
            1.        ],
           ...,
           [0.        , 1.        , 0.11764706, ..., 0.        , 0.        ,
            1.        ],
           [1.        , 0.33333333, 0.60294118, ..., 0.        , 1.        ,
            1.        ],
           [0.        , 0.        , 0.60294118, ..., 0.        , 0.        ,
            1.        ]])
    ```

5. Create a variable called **label** that contains the data only from the first column (this will be our response variable):

    ```
    label = data[:, 0]
    ```

6. Create a variable called **features** that contains all the data except for the first column (which corresponds to the response variable):

```
features = data[:, 1:]
```

7. Import the **train_test_split** method from **sklearn.model_selection**:

```
from sklearn.model_selection import train_test_split
```

8. Split the data into training and testing sets and save the results into four variables called **features_train**, **features_test**, **label_train**, and **label_test**. Use 20% of the data for testing and specify **random_state=7**:

```
features_train, features_test, \
label_train, label_test = train_test_split(features, \
                                           label, \
                                           test_size=0.2, \
                                           random_state=7)
```

9. Import **numpy** as **np**, **tensorflow** as **tf**, and **layers** from **tensorflow.keras**:

```
import numpy as np
import tensorflow as tf
from tensorflow.keras import layers
```

10. Set **1** as the seed for **numpy** and **tensorflow** using **np.random_seed()** and **tf.random.set_seed()**:

```
np.random.seed(1)
tf.random.set_seed(1)
```

11. Instantantiate a **tf.keras.Sequential()** class and save it into a variable called **model**:

```
model = tf.keras.Sequential()
```

12. Instantantiate a **layers.Dense()** class with **16** neurons, **activation='relu'**, and **input_shape=[19]**, then save it into a variable called **layer1**:

```
layer1 = layers.Dense(16, activation='relu', \
                      input_shape=[19])
```

13. Instantiate a second **layers.Dense()** class with **1** neuron and **activation='sigmoid'**, then save it into a variable called **final_layer**:

```
final_layer = layers.Dense(1, activation='sigmoid')
```

14. Add the two layers you just defined to the model using **.add()**:

```
model.add(layer1)
model.add(final_layer)
```

15. Instantiate a **tf.keras.optimizers.Adam()** class with **0.001** as the learning rate and save it into a variable called **optimizer**:

```
optimizer = tf.keras.optimizers.Adam(0.001)
```

16. Compile the neural network using **.compile()** with **loss='binary_crossentropy'**, **optimizer=optimizer, metrics=['accuracy']** as shown in the following code snippet:

```
model.compile(loss='binary_crossentropy', \
              optimizer=optimizer, metrics=['accuracy'])
```

17. Print a summary of the model using **.summary()**:

```
model.summary()
```

The expected output is this:

```
Model: "sequential"

_____
Layer (type)                 Output Shape              Param #
=================================================================
dense (Dense)                (None, 16)                320
_____
dense_1 (Dense)              (None, 1)                 17
=================================================================
Total params: 337
Trainable params: 337
Non-trainable params: 0
_____
```

Figure 6.13: Summary of the sequential model

This output summarizes the architecture of our neural networks. We can see it is composed of three layers, as expected, and we know each layer's output size and number of parameters, which corresponds to the weights and biases. For instance, the first layer has **16** neurons and **320** parameters to be learned (weights and biases).

18. Next, fit the neural networks with the training set and specify **epochs=10**:

```
model.fit(features_train, label_train, epochs=10)
```

The expected output is this:

```
Train on 800 samples
Epoch 1/10
800/800 [==============================] - 0s 349us/sample - loss: 0.7415 - accuracy: 0.3675
Epoch 2/10
800/800 [==============================] - 0s 35us/sample - loss: 0.6817 - accuracy: 0.5612
Epoch 3/10
800/800 [==============================] - 0s 40us/sample - loss: 0.6435 - accuracy: 0.6762
Epoch 4/10
800/800 [==============================] - 0s 35us/sample - loss: 0.6219 - accuracy: 0.6888
Epoch 5/10
800/800 [==============================] - 0s 37us/sample - loss: 0.6125 - accuracy: 0.6888
Epoch 6/10
800/800 [==============================] - 0s 37us/sample - loss: 0.6080 - accuracy: 0.6888
Epoch 7/10
800/800 [==============================] - 0s 38us/sample - loss: 0.6049 - accuracy: 0.6888
Epoch 8/10
800/800 [==============================] - 0s 42us/sample - loss: 0.6025 - accuracy: 0.6888
Epoch 9/10
800/800 [==============================] - 0s 41us/sample - loss: 0.6002 - accuracy: 0.6888
Epoch 10/10
800/800 [==============================] - 0s 32us/sample - loss: 0.5979 - accuracy: 0.6888

<tensorflow.python.keras.callbacks.History at 0x1508c7390>
```

Figure 6.14: Fitting the neural network with the training set

The output provides a lot of information about the training of the neural network. The first line tells us the training set was composed of **800** observations. Then we can see the results of each epoch:

Total processing time in seconds

Processing time by data sample in us/sample

Loss value and accuracy score

The final result of this neural network is the last epoch (**epoch=10**), where we achieved an accuracy score of **0.6888**. But we can see that the trend was improving: the accuracy score was still increasing after each epoch. So, we may get better results if we train the neural network for longer by increasing the number of epochs or lowering the learning rate.

By completing this exercise, you just trained your first classifier. In traditional
machine learning algorithms, you would need to use more lines of code to achieve
this, as you would have to define the entire architecture of the neural network. Here
the neural network got **0.6888** after **10** epochs, but it could still improve if we let it
train for longer. You can try this on your own.

Next, we will be looking at regularization.

REGULARIZATION

As with any machine learning algorithm, neural networks can face the problem of
overfitting when they learn patterns that are only relevant to the training set. In such
a case, the model will not be able to generalize the unseen data.

Luckily, there are multiple techniques that can help reduce the risk of overfitting:

- L1 regularization, which adds a penalty parameter (absolute value of the
  weights) to the loss function

- L2 regularization, which adds a penalty parameter (squared value of the weights)
  to the loss function

- Early stopping, which stops the training if the error for the validation set
  increases while the error decreases for the training set

- Dropout, which will randomly remove some neurons during training

All these techniques can be added at each layer of a neural network we create. We
will be looking at this in the next exercise.

EXERCISE 6.04: PREDICTING BOSTON HOUSE PRICES WITH REGULARIZATION

In this exercise, you will build a neural network that will predict the median house price for a suburb in Boston and see how to add regularizers to a network.

> **NOTE**
>
> The dataset file can also be found in our GitHub repository: https://packt. live/2V9kRUU.
>
> Citation: The data was originally published by *Harrison, D. and Rubinfeld, D.L. 'Hedonic prices and the demand for clean air', J. Environ. Economics & Management, vol.5, 81-102, 1978.*

The dataset is composed of **12** different features that provide information about the suburb and a target variable (**MEDV**). The target variable is numeric and represents the median value of owner-occupied homes in units of $1,000.

The following steps will help you complete the exercise:

1. Open a new Jupyter Notebook file.

2. Import the **pandas** package as **pd**:

```
import pandas as pd
```

3. Create a **file_url** variable containing a link to the raw dataset:

```
file_url = 'https://raw.githubusercontent.com/'\
           'PacktWorkshops/'\
           'The-Applied-Artificial-Intelligence-Workshop'\
           '/master/Datasets/boston_house_price.csv'
```

4. Load the dataset into a variable called **df** using **pd.read_csv()**:

```
df = pd.read_csv(file_url)
```

5. Display the first five rows using **.head()**:

```
df.head()
```

The expected output is this:

	CRIM	ZN	INDUS	CHAS	NOX	RM	AGE	DIS	RAD	TAX	PTRATIO	LSTAT	MEDV
0	0.00632	18.0	2.31	0.0	0.538	6.575	65.2	4.0900	1.0	296.0	15.3	4.98	24.0
1	0.02731	0.0	7.07	0.0	0.469	6.421	78.9	4.9671	2.0	242.0	17.8	9.14	21.6
2	0.02729	0.0	7.07	0.0	0.469	7.185	61.1	4.9671	2.0	242.0	17.8	4.03	34.7
3	0.03237	0.0	2.18	0.0	0.458	6.998	45.8	6.0622	3.0	222.0	18.7	2.94	33.4
4	0.06905	0.0	2.18	0.0	0.458	7.147	54.2	6.0622	3.0	222.0	18.7	5.33	36.2

Figure 6.15: Output showing the first five rows of the dataset

6. Extract the target variable using `.pop()` and save it into a variable called **label**:

```
label = df.pop('MEDV')
```

7. Import the **scale** function from **sklearn.preprocessing**:

```
from sklearn.preprocessing import scale
```

8. Scale the DataFrame, **df**, and save the results into a variable called **scaled_features**. Print its content:

```
scaled_features = scale(df)
scaled_features
```

The expected output is this:

```
array([[-0.41978194,  0.28482986, -1.2879095 , ..., -0.66660821,
        -1.45900038, -1.0755623 ],
       [-0.41733926, -0.48772236, -0.59338101, ..., -0.98732948,
        -0.30309415, -0.49243937],
       [-0.41734159, -0.48772236, -0.59338101, ..., -0.98732948,
        -0.30309415, -1.2087274 ],
       ...,
       [-0.41344658, -0.48772236,  0.11573841, ..., -0.80321172,
         1.17646583, -0.98304761],
       [-0.40776407, -0.48772236,  0.11573841, ..., -0.80321172,
         1.17646583, -0.86530163],
       [-0.41500016, -0.48772236,  0.11573841, ..., -0.80321172,
```

In the output, you can see that all our features are now standardized.

9. Import **train_test_split** from **sklearn.model_selection**:

```
from sklearn.model_selection import train_test_split
```

10. Split the data into training and testing sets and save the results into four variables called **features_train**, **features_test**, **label_train**, and **label_test**. Use 10% of the data for testing and specify **random_state=8**:

```
features_train, features_test, \
label_train, label_test = train_test_split(scaled_features, \
                                           label, \
                                           test_size=0.1, \
                                           random_state=8)
```

11. Import **numpy** as **np**, **tensorflow** as **tf**, and **layers** from **tensorflow.keras**:

```
import numpy as np
import tensorflow as tf
from tensorflow.keras import layers
```

12. Set **8** as the seed for NumPy and TensorFlow using **np.random_seed()** and **tf.random.set_seed()**:

```
np.random.seed(8)
tf.random.set_seed(8)
```

13. Instantiate a **tf.keras.Sequential()** class and save it into a variable called **model**:

```
model = tf.keras.Sequential()
```

14. Next, create a combined **l1** and **l2** regularizer using **tf.keras. regularizers.l1_l2** with **l1=0.01** and **l2=0.01**. Save it into a variable called **regularizer**:

```
regularizer = tf.keras.regularizers.l1_l2(l1=0.1, l2=0.01)
```

15. Instantiate a **layers.Dense()** class with **10** neurons, **activation='relu'**, **input_shape=[12]**, and **kernel_ regularizer=regularizer**, and save it into a variable called **layer1**:

```
layer1 = layers.Dense(10, activation='relu', \
        input_shape=[12], kernel_regularizer=regularizer)
```

16. Instantiate a second **layers.Dense()** class with **1** neuron and save it into a variable called **final_layer**:

```
final_layer = layers.Dense(1)
```

17. Add the two layers you just defined to the model using **.add()** and add a layer in between each of them with **layers.Dropout(0.25)**:

```
model.add(layer1)
model.add(layers.Dropout(0.25))
model.add(final_layer)
```

We added a dropout layer in between each dense layer that will randomly remove 25% of the neurons.

18. Instantiate a **tf.keras.optimizers.SGD()** class with **0.001** as the learning rate and save it into a variable called **optimizer**:

```
optimizer = tf.keras.optimizers.SGD(0.001)
```

19. Compile the neural network using **.compile()** with **loss='mse'**, **optimizer=optimizer, metrics=['mse']**:

```
model.compile(loss='mse', optimizer=optimizer, \
              metrics=['mse'])
```

20. Print a summary of the model using **.summary()**:

```
model.summary()
```

The expected output is this:

```
Model: "sequential"
```

Layer (type)	Output Shape	Param #
dense (Dense)	(None, 10)	130
dropout (Dropout)	(None, 10)	0
dense_1 (Dense)	(None, 1)	11

```
Total params: 141
Trainable params: 141
Non-trainable params: 0
```

Figure 6.16: Summary of the model

This output summarizes the architecture of our neural networks. We can see it is composed of three layers with two dense layers and one dropout layer.

21. Instantiate a **tf.keras.callbacks.EarlyStopping()** class with **monitor='val_loss'** and **patience=2** as the learning rate and save it into a variable called **callback**:

```
callback = tf.keras.callbacks.EarlyStopping(monitor='val_loss', \
                                    patience=2)
```

We just defined a callback stating the neural network will stop its training if the validation loss (**monitor='val_loss'**) does not improve after **2** epochs (**patience=2**).

22. Fit the neural networks with the training set and specify **epochs=50**, **validation_split=0.2**, **callbacks=[callback]**, and **verbose=2**:

```
model.fit(features_train, label_train, \
          epochs=50, validation_split = 0.2, \
          callbacks=[callback], verbose=2)
```

The expected output is this:

```
Epoch 11/50
364/364 - 0s - loss: 70.9273 - mse: 67.1181 - val_loss: 23.7106 - val_mse: 19.9165
Epoch 12/50
364/364 - 0s - loss: 73.2227 - mse: 69.4411 - val_loss: 22.9263 - val_mse: 19.1747
Epoch 13/50
364/364 - 0s - loss: 74.5878 - mse: 70.8375 - val_loss: 21.9391 - val_mse: 18.2101
Epoch 14/50
364/364 - 0s - loss: 72.1759 - mse: 68.4573 - val_loss: 22.3529 - val_mse: 18.6474
Epoch 15/50
364/364 - 0s - loss: 69.3952 - mse: 65.6949 - val_loss: 21.3404 - val_mse: 17.6276
Epoch 16/50
364/364 - 0s - loss: 68.7630 - mse: 65.0606 - val_loss: 20.2952 - val_mse: 16.5971
Epoch 17/50
364/364 - 0s - loss: 69.6016 - mse: 65.9301 - val_loss: 20.4639 - val_mse: 16.8256
Epoch 18/50
364/364 - 0s - loss: 59.4020 - mse: 55.7631 - val_loss: 19.1013 - val_mse: 15.4458
Epoch 19/50
364/364 - 0s - loss: 73.0534 - mse: 69.4134 - val_loss: 18.7573 - val_mse: 15.1291
Epoch 20/50
364/364 - 0s - loss: 57.3614 - mse: 53.7313 - val_loss: 17.2894 - val_mse: 13.6550
Epoch 21/50
364/364 - 0s - loss: 69.8086 - mse: 66.1753 - val_loss: 17.7770 - val_mse: 14.1564
Epoch 22/50
364/364 - 0s - loss: 56.6582 - mse: 53.0436 - val_loss: 17.6301 - val_mse: 14.0292

<tensorflow.python.keras.callbacks.History at 0x15140dd90>
```

Figure 6.17: Fitting the neural network with the training set

In the output, we see that the neural network stopped its training after the 22nd epoch. It stopped well before the maximum number of epochs, **50**. This is due to the callback we set earlier: if the validation loss does not improve after two epochs, the training should stop.

> **NOTE**
>
> To access the source code for this specific section, please refer to https://packt.live/2Yobbba.
>
> You can also run this example online at https://packt.live/37SVSu6. **You must execute the entire Notebook in order to get the desired result.**

You just applied multiple regularization techniques and trained a neural network to predict the median value of housing in Boston suburbs.

ACTIVITY 6.01: FINDING THE BEST ACCURACY SCORE FOR THE DIGITS DATASET

In this activity, you will be training and evaluating a neural network that will be recognizing handwritten digits from the images provided by the MNIST dataset. You will be focusing on achieving an optimal accuracy score.

> **NOTE**
>
> You can read more about this dataset on TensorFlow's website at https://www.tensorflow.org/datasets/catalog/mnist.
>
> Citation: This dataset was originally shared by *Yann Lecun*.

The following steps will help you complete the activity:

1. Import the MNIST dataset.

2. Standardize the data by applying a division by **255**.

3. Create a neural network architecture with the following layers:

    A flatten input layer using **layers.Flatten(input_shape=(28,28))**

    A fully connected layer with **layers.Dense(128, activation='relu')**

    A dropout layer with **layers.Dropout(0.25)**

A fully connected layer with `layers.Dense(10, activation='softmax')`

4. Specify an **Adam** optimizer with a learning rate of **0.001**.

5. Define an early stopping on the validation loss and patience of **5**.

6. Train the model.

7. Evaluate the model and find the accuracy score.

The expected output is this:

```
Train on 48000 samples, validate on 12000 samples
Epoch 1/10
48000/48000 - 3s - loss: 0.3383 - accuracy: 0.9007 - val_loss: 0.1580 - val_accuracy: 0.9540
Epoch 2/10
48000/48000 - 2s - loss: 0.1666 - accuracy: 0.9509 - val_loss: 0.1235 - val_accuracy: 0.9645
Epoch 3/10
48000/48000 - 2s - loss: 0.1274 - accuracy: 0.9612 - val_loss: 0.1043 - val_accuracy: 0.9706
Epoch 4/10
48000/48000 - 2s - loss: 0.1036 - accuracy: 0.9688 - val_loss: 0.0973 - val_accuracy: 0.9705
Epoch 5/10
48000/48000 - 2s - loss: 0.0877 - accuracy: 0.9731 - val_loss: 0.0832 - val_accuracy: 0.9748
Epoch 6/10
48000/48000 - 2s - loss: 0.0774 - accuracy: 0.9759 - val_loss: 0.0838 - val_accuracy: 0.9752
Epoch 7/10
48000/48000 - 2s - loss: 0.0679 - accuracy: 0.9781 - val_loss: 0.0830 - val_accuracy: 0.9749
Epoch 8/10
48000/48000 - 2s - loss: 0.0637 - accuracy: 0.9795 - val_loss: 0.0813 - val_accuracy: 0.9768
Epoch 9/10
48000/48000 - 2s - loss: 0.0576 - accuracy: 0.9809 - val_loss: 0.0851 - val_accuracy: 0.9766
Epoch 10/10
48000/48000 - 2s - loss: 0.0523 - accuracy: 0.9825 - val_loss: 0.0787 - val_accuracy: 0.9779

<tensorflow.python.keras.callbacks.History at 0x161c8ed50>
```

Figure 6.18: Expected accuracy score

> **NOTE**
>
> The solution for this activity can be found on page 378

In the next part, we will dive into deep learning topics.

DEEP LEARNING

Now that we are comfortable in building and training a neural network with one hidden layer, we can look at more complex architecture with deep learning.

Deep learning is just an extension of traditional neural networks but with deeper and more complex architecture. Deep learning can model very complex patterns, be applied in tasks such as detecting objects in images and translating text into a different language.

SHALLOW VERSUS DEEP NETWORKS

Now that we are comfortable in building and training a neural network with one hidden layer, we can look at more complex architecture with deep learning.

As mentioned earlier, we can add more hidden layers to a neural network. This will increase the number of parameters to be learned but can potentially help to model more complex patterns. This is what deep learning is about: increasing the depth of a neural network to tackle more complex problems.

For instance, we can add a second layer to the neural network we presented earlier in the section on forward propagation and loss functions:

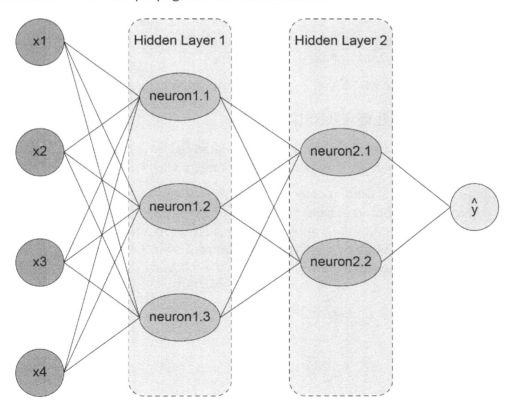

Figure 6.19: Figure showing two hidden layers in a neural network

In theory, we can add an infinite number of hidden layers. But there is a drawback with deeper networks. Increasing the depth will also increase the number of parameters to be optimized. So, the neural network will have to train for longer. So, as good practice, it is better to start with a simpler architecture and then steadily increase its depth.

COMPUTER VISION AND IMAGE CLASSIFICATION

Deep learning has achieved amazing results in computer vision and natural language processing. Computer vision is a field that involves analyzing digital images. A digital image is a matrix composed of **pixels**. Each pixel has a value between **0** and **255** and this value represents the intensity of the pixel. An image can be black and white and have only one channel. But it can also have colors, and in that case, it will have three channels for the colors red, green, and blue. This digital version of an image that can be fed to a deep learning model.

There are multiple applications of computer vision, such as image classification (recognizing the main object in an image), object detection (localizing different objects in an image), and image segmentation (finding the edges of objects in an image). In this book, we will only look at image classification.

In the next section, we will look at a specific type of architecture: CNNs.

CONVOLUTIONAL NEURAL NETWORKS (CNNS)

CNNs are ANNs that are optimized for image-related pattern recognition. CNNs are based on convolutional layers instead of fully connected layers.

A convolutional layer is used to detect patterns in an image with a filter. A filter is just a matrix that is applied to a portion of an input image through a convolutional operation and the output will be another image (also called a feature map) with the highlighted patterns found by the filter. For instance, a simple filter can be one that recognizes vertical lines on a flower, such as for the following image:

Original Image   Vertical Edge Detection

Figure 6.20: Convolution detecting patterns in an image

These filters are not set in advance but learned by CNNs automatically. After the training is over, a CNN can recognize different shapes in an image. These shapes can be anywhere on the image, and the convolutional operator recognizes similar image information regardless of its exact position and orientation.

CONVOLUTIONAL OPERATIONS

A convolution is a specific type of matrix operation. For an input image, a filter of size **n*n** will go through a specific area of an image and apply an element-wise product and a sum and return the calculated value:

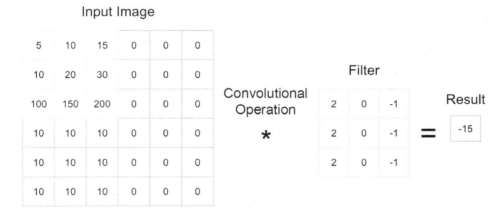

Figure 6.21: Convolutional operations

In the preceding example, we applied a filter to the top-left part of the image. Then we applied an element-wise product that just multiplied an element from the input image to the corresponding value on the filter. In the example, we calculated the following:

- 1st row, 1st column: **5 * 2 = 10**

- 1st row, 2nd column: **10 * 0 = 0**

- 1st row, 3rd column: **15 * (−1) = −15**

- 2nd row, 1st column: **10 * 2 = 20**

- 2nd row, 2nd column: **20 * 0 = 0**

- 2nd row, 3rd column: **30 * (−1) = −30**

- 3rd row, 1st column: **100 * 2 = 200**

- 3rd row, 2nd column: **150 * 0 = 0**

- 3rd row, 3rd column: **200 * (−1) = −200**

Finally, we perform the sum of these values: **10 + 0 -15 + 20 + 0 - 30 + 200 + 0 - 200 = −15**.

Then we will perform the same operation by sliding the filter to the right by one column from the input image. We keep sliding the filter until we have covered the entire image:

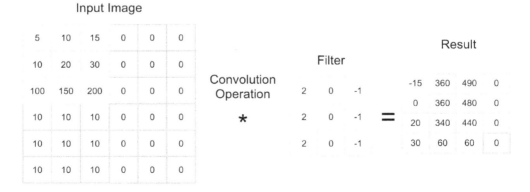

Figure 6.22: Convolutional operations on different rows and columns

Rather than sliding column by column, we can also slide by two, three, or more columns. The parameter defining the length of this sliding operation is called the **stride**.

You may have noticed that the result of the convolutional operation is an image (or feature map) with smaller dimensions than the input image. If you want to keep the exact same dimensions, you can add additional rows and columns with the value 0 around the border of the input image. This operation is called **padding**.

This is what is behind a convolutional operation. A convolutional layer is just the application of this operation with multiple filters.

We can declare a convolutional layer in TensorFlow with the following code snippet:

```
from tensorflow.keras import layers

layers.Conv2D(32, kernel_size=(3, 3), strides=(1,1), \
              padding="valid", activation="relu")
```

In the preceding example, we have instantiated a convolutional layer with **32** filters (also called **kernels**) of size **(3, 3)** with stride of **1** (sliding window by 1 column or row at a time) and no padding (**padding="valid"**).

> **NOTE**
>
> You can read more about this Conv2D class on TensorFlow's website, at https://www.tensorflow.org/api_docs/python/tf/keras/layers/Conv2D.

In TensorFlow, convolutional layers expect the input to be tensors with the following format: (**rows**, **height**, **width**, **channel**). Depending on the dataset, you may have to reshape the images to conform to this requirement. TensorFlow provides a function for this, shown in the following code snippet:

```
features_train.reshape(60000, 28, 28, 1)
```

POOLING LAYER

Another frequent layer in a CNN's architecture is the pooling layer. We have seen previously that the convolutional layer reduces the size of the image if no padding is added. Is this behavior expected? Why don't we keep the exact same size as for the input image? In general, with CNNs, we tend to reduce the size of the feature maps as we progress through different layers. The main reason for this is that we want to have more and more specific pattern detectors closer to the end of the network.

Closer to the beginning of the network, a CNN will tend to have more generic filters, such as vertical or horizontal line detectors, but as it goes deeper, we would, for example, have filters that can detect a dog's tail or a cat's whiskers if we were training a CNN to recognize cats versus dogs, or the texture of objects if we were classifying images of fruits. Also, having smaller feature maps reduces the risk of false patterns being detected.

By increasing the stride, we can further reduce the size of the output feature map. But there is another way to do this: adding a pooling layer after a convolutional layer. A pooling layer is a matrix of a given size and will apply an aggregation function to each area of the feature map. The most frequent aggregation method is finding the maximum value of a group of pixels:

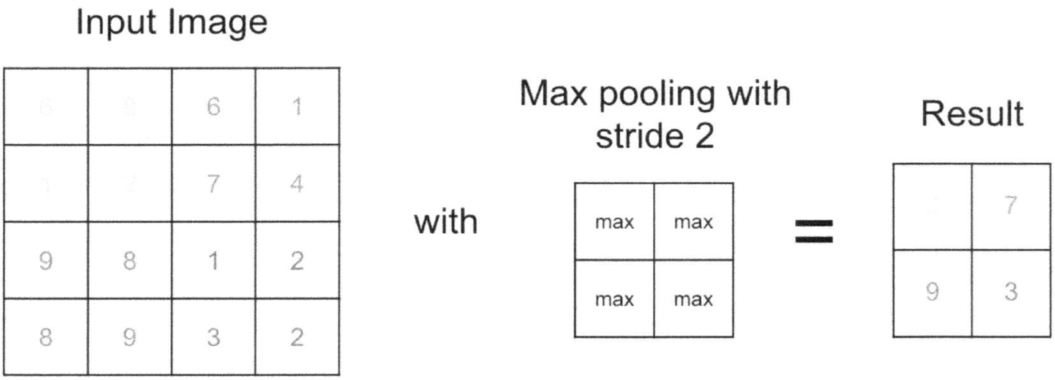

Figure 6.23: Workings of the pooling layer

In the preceding example, we use a max pooling of size (**2, 2**) and **stride=2**. We look at the top-left corner of the feature map and find the maximum value among the pixels **6**, **8**, **1**, and **2** and get the result, **8**. Then we slide the max pooling by a stride of **2** and perform the same operation on the pixels **6**, **1**, **7**, and **4**. We repeat the same operation on the bottom groups and get a new feature map of size (**2, 2**).

In TensorFlow, we can use the **MaxPool2D()** class to declare a max-pooling layer:

```
from tensorflow.keras import layers

layers.MaxPool2D(pool_size=(2, 2), strides=2)
```

> **NOTE**
>
> You can read more about this Conv2D class on TensorFlow's website at
> https://www.tensorflow.org/api_docs/python/tf/keras/layers/MaxPool2D.

CNN ARCHITECTURE

As you saw earlier, you can define your own custom CNN architecture by specifying the type and number of hidden layers, the activation functions to be used, and so on. But this may be a bit daunting for beginners. How do we know how many filters need to be added at each layer or what the right stride will be? We will have to try multiple combinations and see which ones work.

Luckily, a lot of researchers in deep learning have already done such exploratory work and have published the architecture they designed. Currently, the most famous ones are these:

- AlexNet
- VGG
- ResNet
- Inception

> **NOTE**
>
> We will not go through the details of each architecture as it is not in
> the scope of this book, but you can read more about the different CNN
> architectures implemented on TensorFlow at https://www.tensorflow.org/
> api_docs/python/tf/keras/applications.

ACTIVITY 6.02: EVALUATING A FASHION IMAGE RECOGNITION MODEL USING CNNS

In this activity, we will be training a CNN to recognize clothing images that belong to 10 different classes from the Fashion MNIST dataset. We will be finding the accuracy of this CNN model.

> **NOTE**
>
> You can read more about this dataset on TensorFlow's website at https://www.tensorflow.org/datasets/catalog/fashion_mnist.
>
> The original dataset was shared by *Han Xiao*.

The following steps will help you complete the activity:

1. Import the Fashion MNIST dataset.

2. Reshape the training and testing set.

3. Standardize the data by applying a division by **255**.

4. Create a neural network architecture with the following layers:

   Three convolutional layers with **Conv2D(64, (3,3), activation='relu')** followed by **MaxPooling2D(2,2)**

   A flatten layer

   A fully connected layer with **Dense(128, activation=relu)**

   A fully connected layer with **Dense(10, activation='softmax')**

5. Specify an **Adam** optimizer with a learning rate of **0.001**.

6. Train the model.

7. Evaluate the model on the testing set.

The expected output is this:

```
10000/10000 [==============================] - 1s 108us/sample - loss:
0.2746 - accuracy: 0.8976
[0.27461639745235444, 0.8976]
```

> **NOTE**
>
> The solution for this activity can be found on page 382.

In the following section, we will learn about a different type of deep learning architecture: the RNN.

RECURRENT NEURAL NETWORKS (RNNS)

In the last section, we learned how we can use CNNs for computer vision tasks such as classifying images. With deep learning, computers are now capable of achieving and sometimes surpassing human performance. Another field that is attracting a lot of interest from researchers is natural language processing. This is a field where RNNs excel.

In the last few years, we have seen a lot of different applications of RNN technology, such as speech recognition, chatbots, and text translation applications. But RNNs are also quite performant in predicting time series patterns, something that's used for forecasting stock markets.

RNN LAYERS

The common point with all the applications mentioned earlier is that the inputs are sequential. There is a time component with the input. For instance, a sentence is a sequence of words, and the order of words matters; stock market data consists of a sequence of dates with corresponding stock prices.

To accommodate such input, we need neural networks to be able to handle sequences of inputs and be able to maintain an understanding of the relationships between them. One way to do this is to create memory where the network can take into account previous inputs. This is exactly how a basic RNN works:

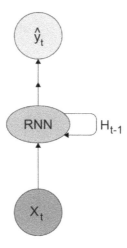

Figure 6.24: Overview of a single RNN

In the preceding figure, we can see a neural network that takes an input called $\mathbf{x}_t$ and performs some transformations and gives the output results, $\widehat{yt}$. Nothing new so far.

But you may have noticed that there is an additional output called H_{t-1} that is an output but also an input to the neural network. This is how RNN simulates memory – by considering its previous results and taking them in as an additional input. Therefore, the result $\widehat{yt}$ will depend on the input x_t but also H_{t-1}. Now, we can represent a sequence of four inputs that get fed into the same neural network:

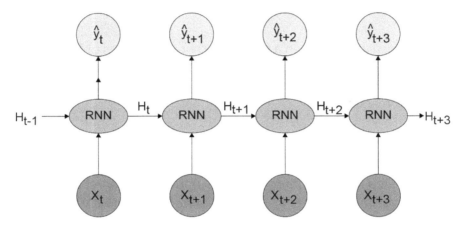

Figure 6.25: Overview of an RNN

We can see the neural network is taking an input (**x**) and generating an output (**y**) at each time step (**t**, **t+1**, ..., **t+3**) but also another output (**h**), which is feeding the next iteration.

> **NOTE**
>
> The preceding figure may be a bit misleading – there is actually only one RNN here (all the RNN boxes in the middle form one neural network), but it is easier to see how the sequencing works in this format.

An RNN cell looks like this on the inside:

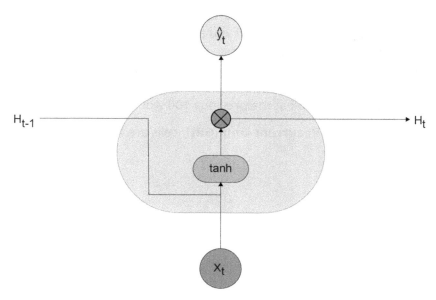

Figure 6.26: Internal workings of an RNN using tanh

It is very similar to a simple neuron, but it takes more inputs and uses **tanh** as the activation function.

> **NOTE**
>
> You can use any activation function in an RNN cell. The default value in TensorFlow is **tanh**.

This is the basic logic of RNNs. In TensorFlow, we can instantiate an RNN layer with **layers.SimpleRNN**:

```
from tensorflow.keras import layers

layers.SimpleRNN(4, activation='tanh')
```

In the code snippet, we created an RNN layer with **4** outputs and the **tanh** activation function (which is the most widely used activation function for RNNs).

THE GRU LAYER

One drawback with the previous type of layer is that the final output takes into consideration all the previous outputs. If you have a sequence of 1,000 input units, the final output, **y**, is influenced by every single previous result. If this sequence was composed of 1,000 words and we were trying to predict the next word, it would really be overkill to have to memorize all of the 1,000 words before making a prediction. Probably, you only need to look at the previous 100 words from the final output.

This is exactly what **Gated Recurrent Unit** (**GRU**) cells are for. Let's look at what is inside them:

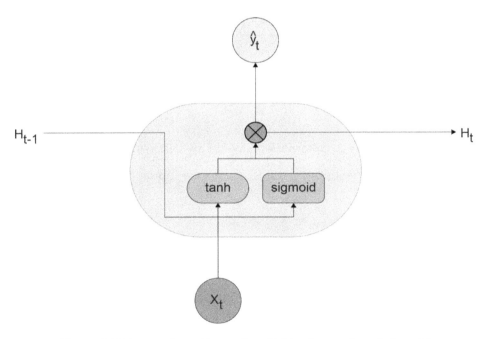

Figure 6.27: Internal workings of an RNN using tanh and sigmoid

Compared to a simple RNN cell, a GRU cell has a few more elements:

- A second activation function, which is `sigmoid`

- A multiplier operation performed before generating the outputs (yt) and H_t

The usual path with `tanh` is still responsible for making a prediction, but this time we will call it the "candidate." The sigmoid path acts as an "update" gate. This will tell the GRU cell whether it needs to discard the use of this candidate or not. Remember that the output ranges between **0** and **1**. If close to 0, the update gate (that is, the sigmoid path) will say we should not consider this candidate.

On the other hand, if it is closer to 1, we should definitely use the result of this candidate.

Remember that the output H_t is related to H_{t-1}, which is related to H_{t-2}, and so on. So, this update gate will also define how much "memory" we should keep. It tends to prioritize previous outputs closer to the current one.

This is the basic logic of GRU (note that the GRU cell has one more component, the reset gate, but for the purpose of simplicity, we will not look at it). In TensorFlow, we can instantiate such a layer with **layers.GRU**:

```
from tensorflow.keras import layers

layers.GRU(4, activation='tanh', \
           recurrent_activation='sigmoid')
```

In the code snippet, we have created a GRU layer with **4** output units and the **tanh** activation function for the candidate prediction and sigmoid for the update gate.

THE LSTM LAYER

There is another very popular type of cell for RNN architecture called the LSTM cell. **LSTM** stands for **Long Short-Term Memory**. LSTM came before GRU, but the latter is much simpler, and this is the reason why we presented it first. Here is what is under the hood of LSTM:

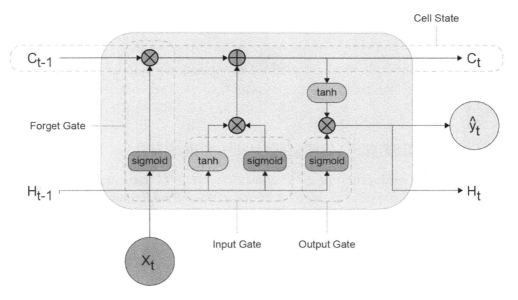

Figure 6.28: Overview of LSTM

At first, this looks very complicated. It is composed of several elements:

- **Cell state**: This is the concatenation of all the previous outputs. It is the "memory" of the LSTM cell.

- **Forget gate**: This is responsible for defining whether we should keep or forget a given memory.

- **Input gate**: This is responsible for defining whether the new memory candidate needs to be updated or not. This new memory candidate is then added to the previous memory.

- **Output gate**: This is responsible for making the prediction based on the previous output (H_{t-1}), the current input (x_t), and the memory.

An LSTM cell can consider previous results but also past memory, and this is the reason why it is so powerful.

In TensorFlow, we can instantiate such a layer with **layers.SimpleRNN**:

```
from tensorflow.keras import layers


layers.LSTM(4, activation='tanh', \
            recurrent_activation='sigmoid')
```

In the code snippet, we have created an LSTM layer with **4** output units and the **tanh** activation function for the candidate prediction and sigmoid for the update gate.

> **NOTE**
>
> You can read more about SimpleRNN implementation in TensorFlow here: https://www.tensorflow.org/api_docs/python/tf/keras/layers/SimpleRNN.

ACTIVITY 6.03: EVALUATING A YAHOO STOCK MODEL WITH AN RNN

In this activity, we will be training an RNN model with LSTM to predict the stock price of Yahoo! based on the data of the past **30** days. We will be finding the optimal mean squared error value and checking whether the model overfits. We will be using the same Yahoo Stock dataset that we saw in *Chapter 2, An Introduction to Regression*.

> **NOTE**
>
> The dataset file can also be found in our GitHub repository: https://packt.live/3fRI5Hk.

The following steps will help you to complete this activity:

1. Import the Yahoo Stock dataset.

2. Extract the **close price** column.

3. Standardize the dataset.

4. Create the previous **30** days' stock price features.

5. Reshape the training and testing sets.

6. Create the neural network architecture with the following layers:

   Five LSTM layers with **LSTM(50, (3,3), activation='relu')** followed by **Dropout(0.2)**

   A fully connected layer with **Dense(1)**

7. Specify an **Adam** optimizer with a learning rate of **0.001**.

8. Train the model.

9. Evaluate the model on the testing set.

The expected output is this:

```
1000/1000 [==============================] - 0s 279us/sample - loss:
0.0016 - mse: 0.0016
[0.00158528157370165, 0.0015852816]
```

> **NOTE**
>
> The solution for this activity can be found on page 387.

In the next section, we will be looking at the hardware needed for deep learning.

HARDWARE FOR DEEP LEARNING

As you may have noticed, training deep learning models takes longer than traditional machine learning algorithms. This is due to the number of calculations required for the forward pass and backpropagation. In this book, we trained very simple models with just a few layers. But there are architectures with hundreds of layers, and some with even more than that. That kind of network can take days or even weeks to train.

To speed up the training process, it is recommended to use a specific piece of hardware called a GPU. GPUs specialize in performing mathematical operations and therefore are perfect for deep learning. Compared to a **Central Processing Unit** (**CPU**), a GPU can be up to 10X faster at training a deep learning model. You can personally buy a GPU and set up your own deep learning computer. You just need to get one that is CUDA-compliant (currently only NVIDIA GPUs are).

Another possibility is to use cloud providers such as AWS or Google Cloud Platform and train your models in the cloud. You will pay only for what you use and can switch them off as soon as you are done. The benefit is that you can scale the configuration up or down depending on the needs of your projects – but be mindful of the cost. You will be charged for the time your instance is up even if you are not training a model. So, don't forget to switch things off if you're not using them.

Finally, Google recently released some new hardware dedicated to deep learning: **Tensor Processing Unit** (**TPUs**). They are much faster than GPUs, but they are quite costly. Currently, only Google Cloud Platform provides such hardware in their cloud instances.

CHALLENGES AND FUTURE TRENDS

As with any new technology, deep learning comes with challenges. One of them is the big barrier to entry. To become a deep learning practitioner, you used to have to know all the mathematical theory behind deep learning very well and be a confirmed programmer. On top of this, you had to learn the specifics of the deep learning framework you chose to use (be it TensorFlow, PyTorch, Caffe, or anything else). For a while, deep learning couldn't reach a broad audience and was mainly limited to researchers. This situation has changed, though it is not perfect. For instance, TensorFlow now comes with a higher-level API called Keras (this is the one you saw in this chapter) that is much easier to use than the core API. Hopefully, this trend will keep going and make deep learning frameworks more accessible to anyone interested in this field.

The second challenge was that deep learning models require a lot of computation power, as mentioned in the previous section. This was again a major blocker for anyone who wanted to have a go at it. Even though the cost of GPUs has gone down, deep learning still requires some upfront investment. Luckily for us, there is now a free option to train deep learning models with GPUs: Google Colab. It is an initiative from Google to promote research by providing temporary cloud computing for free. The only thing you need is a Google account. Once signed up, you can create Notebooks (similar to Jupyter Notebooks) and choose a kernel to be run on a CPU, GPU (limited to 10 hours per day), or even a TPU (limited to ½ hour per day). So, before investing in purchasing or renting out GPU, you can first practice with Google Colab.

> **NOTE**
>
> You can find more information about Google Colab at https://colab.research.google.com/.

More advanced deep learning models can be very deep and require weeks of training. So, it is hard for basic practitioners to use such architecture. But thankfully, a lot of researchers have embraced the open source movement and have shared not only the architectures they have designed but also the weights of the networks. This means you can now access state-of-the-art pre-trained models and fine-tune them to fit your own projects. This is called transfer learning (which is out of the scope of this book). It is very popular in computer vision, where you can find pre-trained models on ImageNet or MS-Coco, for instance, which are large datasets of pictures. Transfer learning is also happening in natural language processing, but it is not as developed as it is for computer vision.

> **NOTE**
>
> You can find more information about these datasets at http://www.image-net.org/ and http://cocodataset.org/.

Another very important topic related to deep learning is the increasing need to be able to interpret model results. Soon, these kinds of algorithms may be regulated, and deep learning practitioners will have to be able to explain why a model is making a given decision. Currently, deep learning models are more like black boxes due to the complexity of the networks. There are already some initiatives from researchers to find ways to interpret and understand deep neural networks, such as *Zeiler and Fergus*, "*Visualizing and Understanding Convolutional Networks*", *ECCV 2014*. However, more work needs to be done in this field with the democratization of such technologies in our day-to-day lives. For instance, we will need to make sure that these algorithms are not biased and are not making unfair decisions affecting specific groups of people.

SUMMARY

We have just completed the entire book of *The Applied Artificial Intelligence Workshop, Second Edition*. In this workshop, we have learned about the fundamentals of AI and its applications. We wrote a Python program to play tic-tac-toe. We learned about search techniques such as breadth-first search and depth-first search and how they can help us solve the tic-tac-toe game.

In the next couple of chapters after that, we learned about supervised learning using regression and classification. These chapters included data preprocessing, train-test splitting, and models that were used in several real-life scenarios. Linear regression, polynomial regression, and support vector machines all came in handy when it came to predicting stock data. Classification was performed using k-nearest neighbor and support vector classifiers. Several activities helped you to apply the basics of classification in an interesting real-life use case: credit scoring.

In *Chapter 4*, *An Introduction to Decision Trees*, you were introduced to decision trees, random forests, and extremely randomized trees. This chapter introduced different means of evaluating the utility of models. We learned how to calculate the accuracy, precision, recall, and F_1 score of models. We also learned how to create the confusion matrix of a model. The models of this chapter were put into practice through the evaluation of car data.

Unsupervised learning was introduced in *Chapter 5, Artificial Intelligence: Clustering*, along with the k-means and hierarchical clustering algorithms. One interesting aspect of these algorithms is that the labels are not given in advance, but they are detected during the clustering process.

This workshop concluded with *Chapter 6, Neural Networks and Deep Learning*, where neural networks and deep learning using TensorFlow was presented. We used these techniques to achieve the best accuracy in real-life applications, such as the detection of written digits, image classification, and time series forecasting.

APPENDIX

CHAPTER 01: INTRODUCTION TO ARTIFICIAL INTELLIGENCE

ACTIVITY 1.01: GENERATING ALL POSSIBLE SEQUENCES OF STEPS IN A TIC-TAC-TOE GAME

Solution:

The following steps will help you to complete this activity:

1. Open a new Jupyter Notebook file.

2. Reuse the function codes of *Steps 2–9* from the previous, *Exercise 1.02, Creating an AI with Random Behavior for the Tic-Tac-Toe Game*.

3. Create a function that maps the **all_moves_from_board_list** function to each element of a list of boards. This way, we will have all of the nodes of a decision tree in each depth:

```
def all_moves_from_board_list(board_list, sign):
    move_list = []
    for board in board_list:
        move_list.extend(all_moves_from_board(board, sign))
    return move_list
```

In the preceding code snippet, we have defined the **all_moves_from_board** function, which will enumerate all the possible moves from the board and add the move to a list called **move_list**.

4. Create a variable called board that contains the **EMPTY_SIGN * 9** decision tree and calls the **all_moves_from_board_list** function with the board and **AI_SIGN**. Save its output in a variable called **all_moves** and print its content:

```
board = EMPTY_SIGN * 9
all_moves = all_moves_from_board(board, AI_SIGN )
all_moves
```

The expected output is this:

```
['X........',
 '.X.......',
 '..X......',
 '...X.....',
 '....X....',
 '.....X...',
 '......X..',
 '.......X.',
 '........X']
```

5. Create a **filter_wins** function that takes the ended games out from the list of moves and appends them in an array containing the board states won by the AI player and the opponent player:

```
def filter_wins(move_list, ai_wins, opponent_wins):
    for board in move_list:
        won_by = game_won_by(board)
        if won_by == AI_SIGN:
            ai_wins.append(board)
            move_list.remove(board)
        elif won_by == OPPONENT_SIGN:
            opponent_wins.append(board)
            move_list.remove(board)
```

In the preceding code snippet, we have defined a **filter_wins** function, which will add the winning state of the board for each player to a list.

6. Use the **count_possibilities** function, which prints and returns the number of decision tree leaves that ended with a draw, that were won by the first player, and that were won by the second player, as shown in the following code snippet:

```
def count_possibilities():
    board = EMPTY_SIGN * 9
    move_list = [board]
    ai_wins = []
    opponent_wins = []
    for i in range(9):
        print('step ' + str(i) + '. Moves: ' \
            + str(len(move_list)))
        sign = AI_SIGN if \
```

```
                    i % 2 == 0 else OPPONENT_SIGN
        move_list = all_moves_from_board_list\
                    (move_list, sign)
        filter_wins(move_list, ai_wins, \
                    opponent_wins)
    print('First player wins: ' + str(len(ai_wins)))
    print('Second player wins: ' + str(len(opponent_wins)))
    print('Draw', str(len(move_list)))
    print('Total', str(len(ai_wins) \
        + len(opponent_wins) + len(move_list)))
    return len(ai_wins), len(opponent_wins), \
        len(move_list), len(ai_wins) \
        + len(opponent_wins) + len(move_list)
```

We have up to **9** steps in each state. In the 0th, 2nd, 4th, 6th, and 8th iterations, the AI player moves. In all the other iterations, the opponent moves. We create all possible moves in all steps and take out the completed games from the move list.

7. Execute the number of possibilities to experience the combinatorial explosion and save the results in four variables called **first_player**, **second_player**, **draw**, and **total**:

```
first_player, second_player, \
draw, total = count_possibilities()
```

The expected output is this:

```
step 0. Moves: 1
step 1. Moves: 9
step 2. Moves: 72
step 3. Moves: 504
step 4. Moves: 3024
step 5. Moves: 13680
step 6. Moves: 49402
step 7. Moves: 111109
step 8. Moves: 156775
First player wins: 106279
Second player wins: 68644
Draw 91150
Total 266073
```

As you can see, the tree of the board states consists of a total of **266073** leaves. The **count_possibilities** function essentially implements a BFS algorithm to traverse all the possible states of the game. Notice that we count these states multiple times because placing an **X** in the top-right corner in *Step 1* and placing an **X** in the top-left corner in *Step 3* leads to similar possible states as starting with the top-left corner and then placing an **X** in the top-right corner. If we implemented the detection of duplicate states, we would have to check fewer nodes. However, at this stage, due to the limited depth of the game, we will omit this step.

A decision tree, however, is identical to the data structure examined by **count_possibilities**. In a decision tree, we explore the utility of each move by investigating all possible future steps up to a certain extent. In our example, we could calculate the utility of the initial moves by observing the number of wins and losses after fixing the first few moves.

> **NOTE**
>
> The root of the tree is the initial state. An internal state of the tree is a state in which a game has not been ended and moves are still possible. A leaf of the tree contains a state where a game has ended.
>
> To access the source code for this specific section, please refer to https://packt.live/3doxPog.
>
> You can also run this example online at https://packt.live/3dpnulz.
>
> You must execute the entire Notebook in order to get the desired result.

ACTIVITY 1.02: TEACHING THE AGENT TO REALIZE SITUATIONS WHEN IT DEFENDS AGAINST LOSSES

Solution:

The following steps will help you to complete this activity:

1. Open a new Jupyter Notebook file.

2. Reuse all the code from *Steps 2–6* from the previous, *Exercise 1.03, Teaching the Agent to Win*.

3. Create a function called **player_can_win** that takes all the moves from the board using the **all_moves_from_board** function and iterates over them using the **next_move** variable.

In each iteration, it checks whether the game can be won by the player.

```
def player_can_win(board, sign):
    next_moves = all_moves_from_board(board, sign)
    for next_move in next_moves:
        if game_won_by(next_move) == sign:
            return True
    return False
```

4. Extend the AI move so that it prefers making safe moves. A move is safe if the opponent cannot win the game in the next step:

```
def ai_move(board):
    new_boards = all_moves_from_board(board, AI_SIGN)
    for new_board in new_boards:
        if game_won_by(new_board) == AI_SIGN:
            return new_board
    safe_moves = []
    for new_board in new_boards:
        if not player_can_win(new_board, OPPONENT_SIGN):
            safe_moves.append(new_board)
    return choice(safe_moves) \
    if len(safe_moves) > 0 else new_boards[0]
```

In the preceding code snippet, we have defined the **ai_move** function, which tells the AI how to move by looking at the list of all the possibilities and choosing one where the player cannot win in the next move. If you test our new application, you will find that the AI has made the correct move.

5. Now, place this logic in the state space generator and check how well the computer player is doing by generating all the possible games:

```
def all_moves_from_board(board, sign):
    move_list = []
    for i, v in enumerate(board):
        if v == EMPTY_SIGN:
            new_board = board[:i] + sign + board[i+1:]
            move_list.append(new_board)
            if game_won_by(new_board) == AI_SIGN:
                return [new_board]
```

```
    if sign == AI_SIGN:
        safe_moves = []
        for move in move_list:
            if not player_can_win(move, OPPONENT_SIGN):
                safe_moves.append(move)
        return safe_moves if len(safe_moves) > 0 else move_list[0:1]
    else:
        return move_list
```

In the preceding code snippet, we have defined a function that generates all possible moves. As soon as we find the next move that can make the player win, we return a move to counter it. We do not care whether the player has multiple options to win the game in one move – we just return the first possibility. If the AI cannot stop the player from winning, we return all possible moves.

Let's see what this means in terms of counting all of the possibilities at each step.

6. Count the options that are possible:

```
first_player, second_player, \
draw, total = count_possibilities()
```

The expected output is this:

```
step 0. Moves: 1
step 1. Moves: 9
step 2. Moves: 72
step 3. Moves: 504
step 4. Moves: 3024
step 5. Moves: 5197
step 6. Moves: 18606
step 7. Moves: 19592
step 8. Moves: 30936
First player wins: 20843
Second player wins: 962
Draw 20243
Total 42048
```

We are doing better than before. We not only got rid of almost 2/3 of possible games again, but, most of the time, the AI player either wins or settles for a draw.

> **NOTE**
>
> To access the source code for this specific section, please refer to https://packt.live/2B0G9xf.
>
> You can also run this example online at https://packt.live/2V7qLpO.
>
> You must execute the entire Notebook in order to get the desired result.

ACTIVITY 1.03: FIXING THE FIRST AND SECOND MOVES OF THE AI TO MAKE IT INVINCIBLE

Solution:

The following steps will help you to complete this activity:

1. Open a new Jupyter Notebook file.

2. Reuse the code from *Steps 2–4* of the previous, *Activity 1.02, Teaching the Agent to Realize Situations When It Defends Against Losses*.

3. Now, count the number of empty fields on the board and make a hardcoded move in case there are 9 or 7 empty fields. You can experiment with different hardcoded moves. We found that occupying any corner, and then occupying the opposite corner, leads to no loss. If the opponent occupies the opposite corner, making a move in the middle results in no losses:

```
def all_moves_from_board(board, sign):
    if sign == AI_SIGN:
        empty_field_count = board.count(EMPTY_SIGN)
        if empty_field_count == 9:
            return [sign + EMPTY_SIGN * 8]
        elif empty_field_count == 7:
            return [board[:8] + sign if board[8] == \
                    EMPTY_SIGN else board[:4] + sign + board[5:]]
    move_list = []
    for i, v in enumerate(board):
        if v == EMPTY_SIGN:
            new_board = board[:i] + sign + board[i+1:]
            move_list.append(new_board)
            if game_won_by(new_board) == AI_SIGN:
```

```
                     return [new_board]
        if sign == AI_SIGN:
            safe_moves = []
            for move in move_list:
                if not player_can_win(move, OPPONENT_SIGN):
                    safe_moves.append(move)
            return safe_moves if len(safe_moves) > 0 else move_list[0:1]
        else:
            return move_list
```

4. Now, verify the state space:

```
first_player, second_player, draw, total = count_possibilities()
```

The expected output is this:

```
step 0. Moves: 1
step 1. Moves: 1
step 2. Moves: 8
step 3. Moves: 8
step 4. Moves: 48
step 5. Moves: 38
step 6. Moves: 108
step 7. Moves: 76
step 8. Moves: 90
First player wins: 128
Second player wins: 0
Draw 60
Total 188
```

After fixing the first two steps, we only need to deal with 8 possibilities instead of 504. We also guided the AI into a state where the hardcoded rules were sufficient enough for it to never lose a game. Fixing the steps is not important because we would give the AI hardcoded steps to start with, but it is important because it is a tool that is used to evaluate and compare each step. After fixing the first two steps, we only need to deal with 8 possibilities instead of 504. We also guided the AI into a state, where the hardcoded rules were sufficient for never losing a game. As you can see, the AI is now nearly invincible and will only win or make a draw.

The best that a player can hope to get against this AI is a draw.

> **NOTE**
>
> To access the source code for this specific section, please refer to https://packt.live/2YnUcpA.
>
> You can also run this example online at https://packt.live/318TBtq.
>
> You must execute the entire Notebook in order to get the desired result.

ACTIVITY 1.04: CONNECT FOUR

Solution:

1. Open a new Jupyter Notebook file.

    Let's set up the **TwoPlayersGame** framework by writing the **init** method.

2. Define the board as a one-dimensional list, like the tic-tac-toe example. We could use a two-dimensional list, too, but modeling will not get much easier or harder. Beyond making initialization like we did in the tic-tac-toe game, we will work a bit further ahead. We will generate all of the possible winning combinations in the game and save them for future use, as shown in the following code snippet:

```
from easyAI import TwoPlayersGame, Human_Player

class ConnectFour(TwoPlayersGame):
    def __init__(self, players):
        self.players = players
        self.board = [0 for i in range(42)]
        self.nplayer = 1
        def generate_winning_tuples():
            tuples = []
            # horizontal
            tuples += [list(range(row*7+column, \
                        row*7+column+4, 1)) \
                        for row in range(6) \
                        for column in range(4)]
            # vertical
            tuples += [list(range(row*7+column, \
                        row*7+column+28, 7)) \
```

```
                       for row in range(3) \
                       for column in range(7)]
            # diagonal forward
            tuples += [list(range(row*7+column, \
                       row*7+column+32, 8)) \
                       for row in range(3) \
                       for column in range(4)]
            # diagonal backward
            tuples += [list(range(row*7+column, \
                       row*7+column+24, 6)) \
                       for row in range(3) \
                       for column in range(3, 7, 1)]
            return tuples
        self.tuples = generate_winning_tuples()
```

3. Next, handle the **possible_moves** function, which is a simple enumeration. Notice that we are using column indices from **1** to **7** in the move names because it is more convenient to start a column indexing with **1** in the human player interface than with zero. For each column, we check whether there is an unoccupied field. If there is one, we will make the column a possible move:

```
def possible_moves(self):
    return [column+1 \
            for column in range(7) \
            if any([self.board[column+row*7] == 0 \
                for row in range(6)])
            ]
```

4. Making a move is like the **possible_moves** function. We check the column of the move and find the first empty cell starting from the bottom. Once we find it, we occupy it. You can also read the implementation of the both the **make_move** function: **unmake_move**. In the **unmake_move** function, we check the column from top to down, and we remove the move at the first non-empty cell. Notice that we rely on the internal representation of **easyAI** so that it does not undo moves that it hasn't made. Otherwise, this function would remove a token of the other player without checking whose token was removed:

```
def make_move(self, move):
    column = int(move) - 1
    for row in range(5, -1, -1):
        index = column + row*7
        if self.board[index] == 0:
```

```
        self.board[index] = self.nplayer
        return


    # optional method (speeds up the AI)
    def unmake_move(self, move):
        column = int(move) - 1
        for row in range(6):
            index = column + row*7
            if self.board[index] != 0:
                self.board[index] = 0
                return
```

5. Since we already have the tuples that we must check, we can mostly reuse the **lose** function from the tic-tac-toe example:

```
    def lose(self):
        return any([all([(self.board[c] == self.nopponent)
                         for c in line])
                   for line in self.tuples])


    def is_over(self):
        return (self.possible_moves() == []) or self.lose()
```

6. Our final task is to implement the **show** method, which prints the board. We will reuse the tic-tac-toe implementation and just change the **show** and **scoring** variables:

```
    def show(self):
        print('\n'+'\n'.join([
            ' '.join([['.', 'O', 'X']\
                     [self.board[7*row+column]] \
                     for column in range(7)])
            for row in range(6)]))


    def scoring(self):
        return -100 if self.lose() else 0


if __name__ == "__main__":
    from easyAI import AI_Player, Negamax
    ai_algo = Negamax(6)
    ConnectFour([Human_Player(), \
                AI_Player(ai_algo)]).play()
```

7. Now that all the functions are complete, you can try out the example. Feel free to play a round or two against your opponent.

   The expected output is this:

```
Player 1 what do you play ? 5

Move #9: player 1 plays 5 :

. . . . . . .
. . . . . . .
X . . . . . .
X . . . . . .
X . . O . . .
O O O X O . .

Move #10: player 2 plays 1 :

. . . . . . .
X . . . . . .
X . . . . . .
X . . . . . .
X . . O . . .
O O O X O . .
```

Figure 1.30: Expected output for the Connect Four game

By completing this activity, you have seen that the opponent is not perfect, but that it plays reasonably well. If you have a strong computer, you can increase the parameter of the **Negamax** algorithm. We encourage you to come up with a better heuristic.

> **NOTE**
>
> To access the source code for this specific section, please refer to https://packt.live/3esk2hl.
>
> You can also run this example online at https://packt.live/3dnkfS5.
>
> You must execute the entire Notebook in order to get the desired result.

CHAPTER 02: AN INTRODUCTION TO REGRESSION

ACTIVITY 2.01: BOSTON HOUSE PRICE PREDICTION WITH POLYNOMIAL REGRESSION OF DEGREES 1, 2, AND 3 ON MULTIPLE VARIABLES

Solution:

1. Open a Jupyter Notebook.

2. Import the required packages and load the Boston House Prices data from **sklearn** into a DataFrame:

```
import numpy as np
import pandas as pd
from sklearn import preprocessing
from sklearn import model_selection
from sklearn import linear_model
from sklearn.preprocessing import PolynomialFeatures
file_url = 'https://raw.githubusercontent.com/'\
            'PacktWorkshops/'\
            'The-Applied-Artificial-Intelligence-Workshop/'\
            'master/Datasets/boston_house_price.csv'
df = pd.read_csv(file_url)
```

The output of **df** is as follows:

	CRIM	ZN	INDUS	CHAS	NOX	RM	AGE	DIS	RAD	TAX	PTRATIO	LSTAT	MEDV
0	0.00632	18.0	2.31	0.0	0.538	6.575	65.2	4.0900	1.0	296.0	15.3	4.98	24.0
1	0.02731	0.0	7.07	0.0	0.469	6.421	78.9	4.9671	2.0	242.0	17.8	9.14	21.6
2	0.02729	0.0	7.07	0.0	0.469	7.185	61.1	4.9671	2.0	242.0	17.8	4.03	34.7
3	0.03237	0.0	2.18	0.0	0.458	6.998	45.8	6.0622	3.0	222.0	18.7	2.94	33.4
4	0.06905	0.0	2.18	0.0	0.458	7.147	54.2	6.0622	3.0	222.0	18.7	5.33	36.2
...	...	...	...	...	...	...	...	...	...	...	...	...	...
501	0.06263	0.0	11.93	0.0	0.573	6.593	69.1	2.4786	1.0	273.0	21.0	9.67	22.4
502	0.04527	0.0	11.93	0.0	0.573	6.120	76.7	2.2875	1.0	273.0	21.0	9.08	20.6
503	0.06076	0.0	11.93	0.0	0.573	6.976	91.0	2.1675	1.0	273.0	21.0	5.64	23.9
504	0.10959	0.0	11.93	0.0	0.573	6.794	89.3	2.3889	1.0	273.0	21.0	6.48	22.0
505	0.04741	0.0	11.93	0.0	0.573	6.030	80.8	2.5050	1.0	273.0	21.0	7.88	11.9

506 rows × 13 columns

Figure 2.28: Output displaying the dataset

Earlier in this chapter, you learned that most of the required packages to perform linear regression come from **sklearn**. We need to import the **preprocessing** module to scale the data, the **linear_model** module to train linear regression, the **PolynomialFeatures** module to transform the inputs for the polynomial regression, and the **model_selection** module to evaluate the performance of each model.

3. Prepare the dataset for prediction by converting the label and features into NumPy arrays and scaling the features:

```
features = np.array(df.drop('MEDV', 1))
label = np.array(df['MEDV'])
scaled_features = preprocessing.scale(features)
```

The output for **features** is as follows:

```
array([[6.3200e-03, 1.8000e+01, 2.3100e+00, ..., 2.9600e+02, 1.5300e+01,
        4.9800e+00],
       [2.7310e-02, 0.0000e+00, 7.0700e+00, ..., 2.4200e+02, 1.7800e+01,
        9.1400e+00],
       [2.7290e-02, 0.0000e+00, 7.0700e+00, ..., 2.4200e+02, 1.7800e+01,
        4.0300e+00],
       ...,
       [6.0760e-02, 0.0000e+00, 1.1930e+01, ..., 2.7300e+02, 2.1000e+01,
        5.6400e+00],
       [1.0959e-01, 0.0000e+00, 1.1930e+01, ..., 2.7300e+02, 2.1000e+01,
        6.4800e+00],
       [4.7410e-02, 0.0000e+00, 1.1930e+01, ..., 2.7300e+02, 2.1000e+01,
        7.8800e+00]])
```

Figure 2.29: Labels and features converted to NumPy arrays

As you can see, our features have been converted into a NumPy array.

The output for the **label** is as follows:

```
array([24. , 21.6, 34.7, 33.4, 36.2, 28.7, 22.9, 27.1, 16.5, 18.9, 15. ,
       18.9, 21.7, 20.4, 18.2, 19.9, 23.1, 17.5, 20.2, 18.2, 13.6, 19.6,
       15.2, 14.5, 15.6, 13.9, 16.6, 14.8, 18.4, 21. , 12.7, 14.5, 13.2,
       13.1, 13.5, 18.9, 20. , 21. , 24.7, 30.8, 34.9, 26.6, 25.3, 24.7,
       21.2, 19.3, 20. , 16.6, 14.4, 19.4, 19.7, 20.5, 25. , 23.4, 18.9,
       35.4, 24.7, 31.6, 23.3, 19.6, 18.7, 16. , 22.2, 25. , 33. , 23.5,
       19.4, 22. , 17.4, 20.9, 24.2, 21.7, 22.8, 23.4, 24.1, 21.4, 20. ,
       20.8, 21.2, 20.3, 28. , 23.9, 24.8, 22.9, 23.9, 26.6, 22.5, 22.2,
       23.6, 28.7, 22.6, 22. , 22.9, 25. , 20.6, 28.4, 21.4, 38.7, 43.8,
       33.2, 27.5, 26.5, 18.6, 19.3, 20.1, 19.5, 19.5, 20.4, 19.8, 19.4,
       21.7, 22.8, 18.8, 18.7, 18.5, 18.3, 21.2, 19.2, 20.4, 19.3, 22. ,
       20.3, 20.5, 17.3, 18.8, 21.4, 15.7, 16.2, 18. , 14.3, 19.2, 19.6,
       23. , 18.4, 15.6, 18.1, 17.4, 17.1, 13.3, 17.8, 14. , 14.4, 13.4,
       15.6, 11.8, 13.8, 15.6, 14.6, 17.8, 15.4, 21.5, 19.6, 15.3, 19.4,
       17. , 15.6, 13.1, 41.3, 24.3, 23.3, 27. , 50. , 50. , 50. , 22.7,
       25. , 50. , 23.8, 23.8, 22.3, 17.4, 19.1, 23.1, 23.6, 22.6, 29.4,
       23.2, 24.6, 29.9, 37.2, 39.8, 36.2, 37.9, 32.5, 26.4, 29.6, 50. ,
       32. , 29.8, 34.9, 37. , 30.5, 36.4, 31.1, 29.1, 50. , 33.3, 30.3,
       34.6, 34.9, 32.9, 24.1, 42.3, 48.5, 50. , 22.6, 24.4, 22.5, 24.4,
       20. , 21.7, 19.3, 22.4, 28.1, 23.7, 25. , 23.3, 28.7, 21.5, 23. ,
```

Figure 2.30: Output showing the expected labels

As you can see, our labels have been converted into a NumPy array.

The output for **scaled_features** is as follows:

```
array([[-0.41978194,  0.28482986, -1.2879095 , ...,
        -0.66660821, -1.45900038, -1.0755623 ],
       [-0.41733926, -0.48772236, -0.59338101, ...,
        -0.98732948, -0.30309415, -0.49243937],
       [-0.41734159, -0.48772236, -0.59338101, ...,
        -0.98732948, -0.30309415, -1.2087274 ],
       ...,
       [-0.41344658, -0.48772236,  0.11573841, ...,
        -0.80321172,  1.17646583, -0.98304761],
       [-0.40776407, -0.48772236,  0.11573841, ...,
        -0.80321172,  1.17646583, -0.86530163],
       [-0.41500016, -0.48772236,  0.11573841, ...,
        -0.80321172,  1.17646583, -0.66905833]])
```

As you can see, our features have been properly scaled.

As we don't have any missing values and we are not trying to predict a future value as we did in *Exercise 2.03, Preparing the Quandl Data for Prediction*, we can directly convert the label (**'MEDV'**) and features into NumPy arrays. Then, we can scale the arrays of features using the **preprocessing.scale()** function.

4. Create three different set of features by transforming the scaled features into a suitable format for each of the polynomial regressions:

```
poly_1_scaled_features = PolynomialFeatures(degree=1)\
                         .fit_transform(scaled_features)
poly_2_scaled_features = PolynomialFeatures(degree=2)\
                         .fit_transform(scaled_features)
poly_3_scaled_features = PolynomialFeatures(degree=3)\
                         .fit_transform(scaled_features)
```

The output for **poly_1_scaled_features** is as follows:

```
array([[ 1.        , -0.41978194,  0.28482986, ..., -0.66660821,
        -1.45900038, -1.0755623 ],
       [ 1.        , -0.41733926, -0.48772236, ..., -0.98732948,
        -0.30309415, -0.49243937],
       [ 1.        , -0.41734159, -0.48772236, ..., -0.98732948,
        -0.30309415, -1.2087274 ],
       ...,
       [ 1.        , -0.41344658, -0.48772236, ..., -0.80321172,
         1.17646583, -0.98304761],
       [ 1.        , -0.40776407, -0.48772236, ..., -0.80321172,
         1.17646583, -0.86530163],
       [ 1.        , -0.41500016, -0.48772236, ..., -0.80321172,
         1.17646583, -0.66905833]])
```

Our **scaled_features** variable has been properly transformed for the polynomial regression of degree **1**.

The output for **poly_2_scaled_features** is as follows:

```
array([[ 1.        , -0.41978194,  0.28482986, ...,  2.12868211,
         1.56924581,  1.15683427],
       [ 1.        , -0.41733926, -0.48772236, ...,  0.09186606,
         0.14925549,  0.24249653],
       [ 1.        , -0.41734159, -0.48772236, ...,  0.09186606,
         0.3663582 ,  1.46102192],
       ...,
       [ 1.        , -0.41344658, -0.48772236, ...,  1.38407185,
        -1.15652192,  0.9663826 ],
       [ 1.        , -0.40776407, -0.48772236, ...,  1.38407185,
        -1.0179978 ,  0.74874691],
       [ 1.        , -0.41500016, -0.48772236, ...,  1.38407185,
        -0.78712427,  0.44763905]])
```

Figure 2.31: Output showing poly_2_scaled_features

Our **scaled_features** variable has been properly transformed for the polynomial regression of degree **2**.

The output for **poly_3_scaled_features** is as follows:

```
array([[ 1.        , -0.41978194,  0.28482986, ..., -2.28953024,
        -1.68782164, -1.24424733],
       [ 1.        , -0.41733926, -0.48772236, ..., -0.04523847,
        -0.07349928, -0.11941484],
       [ 1.        , -0.41734159, -0.48772236, ..., -0.11104103,
        -0.4428272 , -1.76597723],
       ...,
       [ 1.        , -0.41344658, -0.48772236, ..., -1.36060852,
         1.13691611, -0.9500001 ],
       [ 1.        , -0.40776407, -0.48772236, ..., -1.19763962,
         0.88087515, -0.64789192],
       [ 1.        , -0.41500016, -0.48772236, ..., -0.9260248 ,
         0.52663205, -0.29949664]])
```

Our **scaled_features** variable has been properly transformed for the polynomial regression of degree **3**.

We had to transform the scaled features in three different ways as each degree of polynomial regression required a different input transformation.

5. Split the data into a training set and a testing set with **random state = 8**:

```
(poly_1_features_train, poly_1_features_test, \
poly_label_train, poly_label_test) = \
model_selection.train_test_split(poly_1_scaled_features, \
                                 label, \
                                 test_size=0.1, \
                                 random_state=8)

(poly_2_features_train, poly_2_features_test, \
poly_label_train, poly_label_test) = \
model_selection.train_test_split(poly_2_scaled_features, \
                                 label, \
                                 test_size=0.1, \
                                 random_state=8)

(poly_3_features_train, poly_3_features_test, \
poly_label_train, poly_label_test) = \
model_selection.train_test_split(poly_3_scaled_features, \
                                 label, \
                                 test_size=0.1, \
                                 random_state=8)
```

As we have three different sets of scaled transformed features but the same set of labels, we had to perform three different splits. By using the same set of labels and **random_state** in each splitting, we ensure that we obtain the same **poly_label_train** and **poly_label_test** for every split.

6. Perform a polynomial regression of degree 1 and evaluate whether the model is overfitting:

```
model_1 = linear_model.LinearRegression()
model_1.fit(poly_1_features_train, poly_label_train)
model_1_score_train = model_1.score(poly_1_features_train, \
                                     poly_label_train)
model_1_score_test = model_1.score(poly_1_features_test, \
                                    poly_label_test)
```

The output for **model_1_score_train** is as follows:

```
0.7406006443486721
```

The output for **model_1_score_test** is as follows:

```
0.6772229017901507
```

To estimate whether a model is overfitting or not, we need to compare the scores of the model applied to the training set and testing set. If the score for the training set is much higher than the test set, we are overfitting. This is the case here where the polynomial regression of degree 1 achieved a score of **0.74** for the training set compared to **0.68** for the testing set.

7. Perform a polynomial regression of degree 2 and evaluate whether the model is overfitting:

```
model_2 = linear_model.LinearRegression()
model_2.fit(poly_2_features_train, poly_label_train)
model_2_score_train = model_2.score(poly_2_features_train, \
                                    poly_label_train)
model_2_score_test = model_2.score(poly_2_features_test, \
                                   poly_label_test)
```

The output for **model_2_score_train** is as follows:

```
0.9251199698832675
```

The output for **model_2_score_test** is as follows:

```
0.8253870684280571
```

Like with the polynomial regression of degree 1, our polynomial regression of degree 2 is overfitting even more than degree 1, but has managed to achieve better results at the end.

8. Perform a polynomial regression of degree 3 and evaluate whether the model is overfitting:

```
model_3 = linear_model.LinearRegression()
model_3.fit(poly_3_features_train, poly_label_train)
model_3_score_train = model_3.score(poly_3_features_train, \
                                    poly_label_train)
model_3_score_test = model_3.score(poly_3_features_test, \
                                   poly_label_test)
```

The output for **model_3_score_train** is as follows:

```
0.9910498071894897
```

The output for **model_3_score_test** is as follows:

```
-8430.781888645262
```

These results are very interesting because the polynomial regression of degree 3 managed to achieve a near-perfect score with **0.99** (1 is the maximum). This is a warning sign that our model is overfitting too much. We have the confirmation of this warning when the model is applied to the testing set and achieves a very low negative score of **-8430**. As a reminder, a score of 0 can be achieved by using the mean of the data as a prediction. This means that our third model managed to make worse predictions than just using the mean.

9. Compare the predictions of the 3 models against the label on the testing set:

```
model_1_prediction = model_1.predict(poly_1_features_test)
model_2_prediction = model_2.predict(poly_2_features_test)
model_3_prediction = model_3.predict(poly_3_features_test)

df_prediction = pd.DataFrame(poly_label_test)
df_prediction.rename(columns = {0:'label'}, inplace = True)
df_prediction['model_1_prediction'] = \
pd.DataFrame(model_1_prediction)
df_prediction['model_2_prediction'] = \
pd.DataFrame(model_2_prediction)
df_prediction['model_3_prediction'] = \
pd.DataFrame(model_3_prediction)
```

The output of **df_prediction** is as follows:

	label	model_1_prediction	model_2_prediction	model_3_prediction
0	18.5	19.269554	19.885620	21.067408
1	12.7	11.434612	14.470337	11.703696
2	21.9	37.610026	32.721497	-3713.860431
3	22.0	26.985628	26.337830	1.711389
4	50.0	40.548986	45.107178	-448.022868
5	36.2	27.730936	26.509888	45.594573
6	16.5	10.601832	18.077148	-6.973548
7	32.4	36.478511	32.267456	1584.038370
8	24.6	29.094356	27.081970	25.674404
9	50.0	34.409453	54.093201	-285.162605
10	13.9	14.138314	15.555481	14.549911
11	11.9	7.143616	14.295105	22.325523

Figure 2.32: Output showing the expected predicted values

After applying the **predict** function for each model on their respective testing set, in order to get the predicted values, we convert them into a single **df_prediction** DataFrame with the label values. Increasing the number of degrees in polynomial regressions does not necessarily mean that the model will perform better compared to one with a lower degree. In fact, increasing the degree will lead to more overfitting on the training data.

> **NOTE**
>
> To access the source code for this specific section, please refer to https://packt.live/3eD8gAY.
>
> You can also run this example online at https://packt.live/3etadjp.
>
> You must execute the entire Notebook in order to get the desired result.

In this activity, we learned how to perform polynomial regressions of degrees 1 to 3 with multiple variables on the Boston House Price dataset and saw how increasing the degrees led to overfitted models.

CHAPTER 03: AN INTRODUCTION TO CLASSIFICATION

ACTIVITY 3.01: INCREASING THE ACCURACY OF CREDIT SCORING

Solution:

1. Open a new Jupyter Notebook file and execute all the steps from the previous exercise, *Exercise 3.04*, *K-Nearest Neighbors Classification in Scikit-Learn*.

2. Import **neighbors** from **sklearn**:

```
from sklearn import neighbors
```

3. Create a function called **fit_knn** that takes the following parameters: **k**, **p**, **features_train**, **label_train**, **features_test**, and **label_test**. This function will fit **KNeighborsClassifier** with the training set and print the accuracy score for the training and testing sets, as shown in the following code snippet:

```
def fit_knn(k, p, features_train, label_train, \
            features_test, label_test):
    classifier = neighbors.KNeighborsClassifier(n_neighbors=k, p=p)
    classifier.fit(features_train, label_train)
    return classifier.score(features_train, label_train), \
           classifier.score(features_test, label_test)
```

4. Call the **fit_knn()** function with **k=5** and **p=2**, save the results in **2** variables, and print them. These variables are **acc_train_1** and **acc_test_1**:

```
acc_train_1, acc_test_1 = fit_knn(5, 2, features_train, \
                                  label_train, \
                                  features_test, label_test)
acc_train_1, acc_test_1
```

The expected output is this:

```
(0.78625, 0.75)
```

With **k=5** and **p=2**, KNN achieved a good accuracy score close to **0.78**. But the score is quite different from the training and testing sets, which means the model is overfitting.

5. Call the **fit_knn()** function with **k=10** and **p=2**, save the results in **2** variables, and print them. These variables are **acc_train_2** and **acc_test_2**:

```
acc_train_2, acc_test_2 = fit_knn(10, 2, features_train, \
                                  label_train, \
                                  features_test, label_test)
acc_train_2, acc_test_2
```

The expected output is this:

```
(0.775, 0.785)
```

Increasing the number of neighbors to 10 has decreased the accuracy score of the training set, but now it is very close to the testing set.

6. Call the **fit_knn()** function with **k=15** and **p=2**, save the results in **2** variables, and print them. These variables are **acc_train_3** and **acc_test_3**:

```
acc_train_3, acc_test_3 = fit_knn(15, 2, features_train, \
                                  label_train, \
                                  features_test, label_test)
acc_train_3, acc_test_3
```

The expected output is this:

```
(0.76625, 0.79)
```

With **k=15** and **p=2**, the difference between the training and testing sets has increased.

7. Call the **fit_knn()** function with **k=25** and **p=2**, save the results in **2** variables, and print them. These variables are **acc_train_4** and **acc_test_4**:

```
acc_train_4, acc_test_4 = fit_knn(25, 2, features_train, \
                                  label_train, \
                                  features_test, label_test)
acc_train_4, acc_test_4
```

The expected output is this:

```
(0.7375, 0.77)
```

Increasing the number of neighbors to **25** has a significant impact on the training set. However, the model is still overfitting.

8. Call the **fit_knn()** function with **k=50** and **p=2**, save the results in **2** variables, and print them. These variables are **acc_train_5** and **acc_test_5**:

```
acc_train_5, acc_test_5 = fit_knn(50, 2, features_train, \
                                  label_train, \
                                  features_test, label_test)
acc_train_5, acc_test_5
```

The expected output is this:

```
(0.70625, 0.775)
```

Bringing the number of neighbors to **50** neither improved the model's performance or the overfitting issue.

9. Call the **fit_knn()** function with **k=5** and **p=1**, save the results in **2** variables, and print them. These variables are **acc_train_6** and **acc_test_6**:

```
acc_train_6, acc_test_6 = fit_knn(5, 1, features_train, \
                                  label_train, \
                                  features_test, label_test)
acc_train_6, acc_test_6
```

The expected output is this:

```
(0.8, 0.735)
```

Changing to the Manhattan distance has helped increase the accuracy of the training set, but the model is still overfitting.

10. Call the **fit_knn()** function with **k=10** and **p=1**, save the results in **2** variables, and print them. These variables are **acc_train_7** and **acc_test_7**:

```
acc_train_7, acc_test_7 = fit_knn(10, 1, features_train, \
                                  label_train, \
                                  features_test, label_test)
acc_train_7, acc_test_7
```

The expected output is this:

```
(0.77, 0.785)
```

With **k=10**, the accuracy score for the training and testing sets are quite close to each other: around **0.78**.

11. Call the **fit_knn()** function with **k=15** and **p=1**, save the results in **2** variables, and print them. These variables are **acc_train_8** and **acc_test_8**:

```
acc_train_8, acc_test_8 = fit_knn(15, 1, features_train, \
                                  label_train, \
                                  features_test, label_test)
acc_train_8, acc_test_8
```

The expected output is this:

```
(0.7575, 0.775)
```

Bumping **k** to **15**, the model achieved a better accuracy score and is not overfitting very much.

12. Call the **fit_knn()** function with **k=25** and **p=1**, save the results in **2** variables, and print them. These variables are **acc_train_9** and **acc_test_9**:

```
acc_train_9, acc_test_9 = fit_knn(25, 1, features_train, \
                                  label_train, \
                                  features_test, label_test)
acc_train_9, acc_test_9
```

The expected output is this:

```
(0.745, 0.8)
```

With **k=25**, the difference between the training and testing sets' accuracy is increasing, so the model is overfitting.

13. Call the **fit_knn()** function with **k=50** and **p=1**, save the results in **2** variables, and print them. These variables are **acc_train_10** and **acc_test_10**:

```
acc_train_10, acc_test_10 = fit_knn(50, 1, features_train, \
                                    label_train, \
                                    features_test, label_test)
acc_train_10, acc_test_10
```

The expected output is this:

```
(0.70875, 0.78)
```

With **k=50**, the model's performance on the training set dropped significantly and the model is definitely overfitting.

In this activity, we tried multiple combinations of hyperparameters for **n_neighbors** and **p**. The best one we found was for **n_neighbors=10** and **p=2**. With these hyperparameters, the model is not overfitting much and it achieved an accuracy score of around **78%** for both the training and testing sets.

> **NOTE**
>
> To access the source code for this specific section, please refer to https://packt.live/2V5TOtG.
>
> You can also run this example online at https://packt.live/2Bx0yd8.
>
> You must execute the entire Notebook in order to get the desired result.

ACTIVITY 3.02: SUPPORT VECTOR MACHINE OPTIMIZATION IN SCIKIT-LEARN

Solution:

1. Open a new Jupyter Notebook file and execute all the steps mentioned in the previous, *Exercise 3.04, K-Nearest Neighbor Classification in scikit-learn*.

2. Import **svm** from **sklearn**:

```
from sklearn import svm
```

3. Create a function called **fit_knn** that takes the following parameters: **features_train**, **label_train**, **features_test**, **label_test**, **kernel="linear"**, **C=1**, **degree=3**, and **gamma='scale'**. This function will fit an SVC with the training set and print the accuracy score for both the training and testing sets:

```
def fit_svm(features_train, label_train, \
            features_test, label_test, \
            kernel="linear", C=1, \
            degree=3, gamma='scale'):
    classifier = svm.SVC(kernel=kernel, C=C, \
                         degree=degree, gamma=gamma)
    classifier.fit(features_train, label_train)
    return classifier.score(features_train, label_train), \
           classifier.score(features_test, label_test)
```

4. Call the **fit_knn()** function with the default hyperparameter values, save the results in **2** variables, and print them. These variables are **acc_train_1** and **acc_test_1**:

```
acc_train_1, \
acc_test_1 =  fit_svm(features_train, \
                        label_train, \
                        features_test, \
                        label_test)
acc_train_1,  acc_test_1
```

The expected output is this:

```
(0.71625, 0.75)
```

With the default hyperparameter values (linear model), the performance of the model is quite different between the training and the testing set.

5. Call the **fit_knn()** function with **kernel="poly"**, **C=1**, **degree=4**, and **gamma=0.05**, save the results in **2** variables, and print them. These variables are **acc_train_2** and **acc_test_2**:

```
acc_train_2, \
acc_test_2 = fit_svm(features_train, label_train, \
                        features_test, label_test, \
                        kernel="poly",   C=1, \
                        degree=4, gamma=0.05)
acc_train_2,  acc_test_2
```

The expected output is this:

```
(0.68875, 0.745)
```

With a fourth-degree polynomial, the model is not performing well on the training set.

6. Call the **fit_knn()** function with **kernel="poly"**, **C=2**, **degree=4**, and **gamma=0.05**, save the results in **2** variables, and print them. These variables are **acc_train_3** and **acc_test_3**:

```
acc_train_3, \
acc_test_3 = fit_svm(features_train, \
                     label_train, features_test, \
                     label_test, kernel="poly", \
                     C=2, degree=4, gamma=0.05)
acc_train_3,   acc_test_3
```

The expected output is this:

```
(0.68875, 0.745)
```

Increasing the regularization parameter, **C**, didn't impact the model's performance at all.

7. Call the **fit_knn()** function with **kernel="poly"**, **C=1**, **degree=4**, and **gamma=0.25**, save the results in **2** variables, and print them. These variables are **acc_train_4** and **acc_test_4**:

```
acc_train_4, \
acc_test_4 = fit_svm(features_train, \
                     label_train, features_test, \
                     label_test, kernel="poly", \
                     C=1, degree=4, gamma=0.25)
acc_train_4,   acc_test_4
```

The expected output is this:

```
(0.84625, 0.775)
```

Increasing the value of gamma to **0.25** has significantly improved the model's performance on the training set. However, the accuracy on the testing set is much lower, so the model is overfitting.

8. Call the `fit_knn()` function with **kernel="poly"**, **C=1**, **degree=4**, and **gamma=0.5**, save the results in **2** variables, and print them. These variables are **acc_train_5** and **acc_test_5**:

```
acc_train_5, \
acc_test_5 = fit_svm(features_train, \
                     label_train, features_test, \
                     label_test, kernel="poly", \
                     C=1, degree=4, gamma=0.5)
acc_train_5,  acc_test_5
```

The expected output is this:

```
(0.9575, 0.73)
```

Increasing the value of gamma to **0.5** has drastically improved the model's performance on the training set, but it is definitely overfitting as the accuracy score on the testing set is much lower.

9. Call the `fit_knn()` function with **kernel="poly"**, **C=1**, **degree=4**, and **gamma=0.16**, save the results in **2** variables, and print them. These variables are **acc_train_6** and **acc_test_6**:

```
acc_train_6, \
acc_test_6 = fit_svm(features_train, label_train, \
                     features_test, label_test, \
                     kernel="poly",  C=1, \
                     degree=4, gamma=0.16)
acc_train_6,  acc_test_6
```

The expected output is this:

```
(0.76375, 0.785)
```

With **gamma=0.16**, the model achieved a better accuracy score than it did for the best KNN model. Both the training and testing sets have a score of around **0.77**.

10. Call the **fit_knn()** function with **kernel="sigmoid"**, save the results in **2** variables, and print them. These variables are **acc_train_7** and **acc_test_7**:

```
acc_train_7, \
acc_test_7 = fit_svm(features_train, label_train, \
                     features_test, label_test, \
                     kernel="sigmoid")
acc_train_7,  acc_test_7
```

The expected output is this:

```
(0.635, 0.66)
```

The sigmoid kernel achieved a low accuracy score.

11. Call the **fit_knn()** function with **kernel="rbf"** and **gamma=0.15**, save the results in **2** variables, and print them. These variables are **acc_train_8** and **acc_test_8**:

```
acc_train_8, \
acc_test_8 = fit_svm(features_train, \
                label_train, features_test, \
                label_test, kernel="rbf", \
                gamma=0.15)
acc_train_8,  acc_test_8
```

The expected output is this:

```
(0.7175, 0.765)
```

The **rbf** kernel achieved a good score with **gamma=0.15**. The model is overfitting a bit, though.

12. Call the **fit_knn()** function with **kernel="rbf"** and **gamma=0.25**, save the results in **2** variables, and print them. These variables are **acc_train_9** and **acc_test_9**:

```
acc_train_9, \
acc_test_9 = fit_svm(features_train, \
                label_train, features_test, \
                label_test, kernel="rbf", \
                gamma=0.25)
acc_train_9,  acc_test_9
```

The expected output is this:

```
(0.74, 0.765)
```

The model performance got better with **gamma=0.25**, but it is still overfitting.

13. Call the **fit_knn()** function with **kernel="rbf"** and **gamma=0.35**, save the results in **2** variables, and print them. These variables are **acc_train_10** and **acc_test_10**:

```
acc_train_10, \
acc_test_10 = fit_svm(features_train, label_train, \
                      features_test, label_test, \
                      kernel="rbf", gamma=0.35)
acc_train_10, acc_test_10
```

The expected output is this:

```
(0.78125, 0.775)
```

With the **rbf** kernel and **gamma=0.35**, we got very similar results for the training and testing sets and the model's performance is higher than the best KNN we trained in the previous activity. This is our best model for the German credit dataset.

> **NOTE**
>
> To access the source code for this specific section, please refer to https://packt.live/3fPZlMQ.
>
> You can also run this example online at https://packt.live/3hVlEm3.
>
> You must execute the entire Notebook in order to get the desired result.

In this activity, we tried different values for the main hyperparameters of the SVM classifier: **kernel**, **gamma**, **C**, and **degrees**. We saw how they affected the model's performance and their tendency to overfit. With trial and error, we finally found the best hyperparameter combination and achieved an accuracy score close to 0.78. This process is called **hyperparameter tuning** and is an important step for any data science project.

CHAPTER 04: AN INTRODUCTION TO DECISION TREES

ACTIVITY 4.01: CAR DATA CLASSIFICATION

Solution:

1. Open a new Jupyter Notebook file.

2. Import the **pandas** package as **pd**:

```
import pandas as pd
```

3. Create a new variable called **file_url** that will contain the URL to the raw dataset:

```
file_url = 'https://raw.githubusercontent.com/'\
           'PacktWorkshops/'\
           'The-Applied-Artificial-Intelligence-Workshop/'\
           'master/Datasets/car.csv'
```

4. Load the data using the **pd.read_csv()** method.:

```
df = pd.read_csv(file_url)
```

5. Print the first five rows of **df**:

```
df.head()
```

The output will be as follows:

	buying	maintenance	doors	persons	luggage_boot	safety	class
0	vhigh	vhigh	2	2	small	low	unacc
1	vhigh	vhigh	2	2	small	med	unacc
2	vhigh	vhigh	2	2	small	high	unacc
3	vhigh	vhigh	2	2	med	low	unacc
4	vhigh	vhigh	2	2	med	med	unacc

Figure 4.13: The first five rows of the dataset

6. Import the **preprocessing** module from **sklearn**:

```
from sklearn import preprocessing
```

7. Create a function called **encode()** that takes a DataFrame and column name as parameters. This function will instantiate **LabelEncoder()**, fit it with the unique value of the column, and transform its data. It will return the transformed column:

```
def encode(data_frame, column):
    label_encoder = preprocessing.LabelEncoder()
    label_encoder.fit(data_frame[column].unique())
    return label_encoder.transform(data_frame[column])
```

8. Create a **for** loop that will iterate through each column of **df** and will encode them with the **encode()** function:

```
for column in df.columns:
    df[column] = encode(df, column)
```

9. Now, print the first five rows of **df**:

```
df.head()
```

The output will be as follows:

	buying	maintenance	doors	persons	luggage_boot	safety	class
0	3	3	0	0	2	1	2
1	3	3	0	0	2	2	2
2	3	3	0	0	2	0	2
3	3	3	0	0	1	1	2
4	3	3	0	0	1	2	2

Figure 4.14: The updated first five rows of the dataset

10. Extract the class column using **.pop()** from pandas and save it in a variable called **label**:

```
label = df.pop('class')
```

11. Import **model_selection** from **sklearn**:

```
from sklearn import model_selection
```

12. Split the dataset into training and testing sets with **test_size=0.1** and **random_state=88**:

```
features_train, features_test, label_train, label_test = \
model_selection.train_test_split(df, label, \
                                 test_size=0.1, \
                                 random_state=88)
```

13. Import **DecisionTreeClassifier** from **sklearn**:

```
from sklearn.tree import DecisionTreeClassifier
```

14. Instantiate **DecisionTreeClassifier()** and save it in a variable called **decision_tree**:

```
decision_tree = DecisionTreeClassifier()
```

15. Fit the decision tree with the training set:

```
decision_tree.fit(features_train, label_train)
```

The output will be as follows:

```
DecisionTreeClassifier(class_weight=None, criterion='gini', max_depth=None,
                max_features=None, max_leaf_nodes=None,
                min_impurity_decrease=0.0, min_impurity_split=None,
                min_samples_leaf=1, min_samples_split=2,
                min_weight_fraction_leaf=0.0, presort=False,
                random_state=None, splitter='best')
```

Figure 4.15: Decision tree fit with the training set

16. Print the score of the decision tree on the testing set:

```
decision_tree.score( features_test, label_test )
```

The output will be as follows:

```
0.953757225433526
```

The decision tree is achieving an accuracy score of **0.95** for our first try. This is remarkable.

17. Import **classification_report** from **sklearn.metrics**:

```
from sklearn.metrics import classification_report
```

18. Print the classification report of the test labels and predictions:

```
print(classification_report(label_test, \
        decision_tree.predict(features_test)))
```

The output will be as follows:

	precision	recall	f1-score	support
0	0.89	0.98	0.93	42
1	0.89	0.89	0.89	9
2	0.99	0.97	0.98	114
3	1.00	0.75	0.86	8
accuracy			0.96	173
macro avg	0.94	0.90	0.92	173
weighted avg	0.96	0.96	0.96	173

Figure 4.16: Output showing the expected classification report

From this classification report, we can see that our model is performing quite well for the precision scores for all four classes. Regarding the recall score, we can see that it didn't perform as well for the last class.

> **NOTE**
>
> To access the source code for this specific section, please refer to https://packt.live/3hQDLtr.
>
> You can also run this example online at https://packt.live/2NkEEML.
>
> You must execute the entire Notebook in order to get the desired result.

By completing this activity, you have prepared the car dataset and trained a decision tree model. You have learned how to get its accuracy score and a classification report so that you can analyze its precision and recall scores.

ACTIVITY 4.02: RANDOM FOREST CLASSIFICATION FOR YOUR CAR RENTAL COMPANY

Solution:

1. Open a Jupyter Notebook.

2. Reuse the code mentioned in *Steps 1 - 4* of *Activity 1, Car Data Classification*.

3. Import **RandomForestClassifier** from **sklearn.ensemble**:

```
from sklearn.ensemble import RandomForestClassifier
```

4. Instantiate a random forest classifier with **n_estimators=100**, **max_depth=6**, and **random_state=168**. Save it to a variable called **random_forest_classifier**:

```
random_forest_classifier = \
RandomForestClassifier(n_estimators=100, \
                       max_depth=6, random_state=168)
```

5. Fit the random forest classifier with the training set:

```
random_forest_classifier.fit(features_train, label_train)
```

The output will be as follows:

```
RandomForestClassifier(bootstrap=True, class_weight=None, criterion='gini',
                       max_depth=6, max_features='auto', max_leaf_nodes=None,
                       min_impurity_decrease=0.0, min_impurity_split=None,
                       min_samples_leaf=1, min_samples_split=2,
                       min_weight_fraction_leaf=0.0, n_estimators=100,
                       n_jobs=None, oob_score=False, random_state=168,
                       verbose=0, warm_start=False)
```

Figure 4.17: Logs of the RandomForest classifier with its hyperparameter values

These are the logs of the **RandomForest** classifier with its hyperparameter values.

6. Make predictions on the testing set using the random forest classifier and save them in a variable called **rf_preds_test**. Print its content:

```
rf_preds_test = random_forest_classifier.fit(features_train, \
                                              label_train)
rf_preds_test
```

The output will be as follows:

```
array([[0, 0, 2, 0, 0, 2, 3, 2, 2, 2, 2, 2, 2, 2, 2, 2, 2, 2, 2, 2, 2, 2,
        0, 2, 2, 3, 2, 2, 2, 2, 0, 2, 0, 2, 2, 2, 2, 2, 2, 2, 2, 2, 2, 2,
        2, 0, 2, 0, 0, 0, 2, 2, 0, 2, 0, 2, 0, 2, 2, 2, 0, 2, 2, 2, 2, 0,
        0, 2, 2, 2, 0, 2, 3, 2, 2, 2, 0, 2, 2, 2, 2, 2, 0, 0, 0, 2, 2, 2,
        2, 2, 2, 2, 0, 2, 0, 0, 2, 2, 0, 2, 2, 2, 0, 0, 2, 0, 2, 2, 2, 2,
        0, 2, 2, 0, 2, 2, 3, 0, 2, 2, 2, 2, 0, 0, 2, 2, 2, 0, 2, 2, 2, 0,
        2, 2, 2, 2, 2, 2, 2, 0, 3, 3, 2, 0, 0, 2, 2, 2, 0, 0, 2, 2, 0, 2,
        2, 2, 2, 2, 0, 0, 2, 2, 2, 0, 2, 2, 2, 2, 0, 2, 0, 0, 2]])
```

Figure 4.18: Output showing the predictions on the testing set

7. Import **classification_report** from **sklearn.metrics**:

```
from sklearn.metrics import classification_report
```

8. Print the classification report with the labels and predictions from the test set:

```
print(classification_report(label_test, rf_preds_test))
```

The output will be as follows:

	precision	recall	f1-score	support
0	0.67	0.76	0.71	42
1	0.00	0.00	0.00	9
2	0.92	0.96	0.94	114
3	0.83	0.62	0.71	8
accuracy			0.84	173
macro avg	0.60	0.59	0.59	173
weighted avg	0.80	0.84	0.82	173

Figure 4.19: Output showing the classification report
with the labels and predictions from the test set

The F_1 score in the preceding report shows us that the random forest is performing well on class **2** but not as good for classes **0** and **3**. The model is unable to predict accurately for class **1**, but there were only 9 observations in the testing set. The accuracy score is **0.84**, while the F_1 score is **0.82**.

9. Import **confusion_matrix** from **sklearn.metrics**:

```
from sklearn.metrics import confusion_matrix
```

10. Display the confusion matrix on the true and predicted labels of the testing set:

```
confusion_matrix(label_test, rf_preds_test)
```

The output will be as follows:

```
array([[ 32,  0,  10,  0],
       [  8,  0,   0,  1],
       [  5,  0, 109,  0],
       [  3,  0,   0,  5]])
```

From this confusion matrix, we can see that the **RandomForest** model is having difficulties accurately predicting the first class. It incorrectly predicted 16 cases (8 + 5 + 3) for this class.

11. Print the feature importance score of the test set using **.feature_importance_** and save the results in a variable called **rf_varimp**. Print its contents:

```
rf_varimp = random_forest_classifier.feature_importances_
rf_varimp
```

The output will be as follows:

```
array([0.12676384, 0.10366314, 0.02119621, 0.35266673,
       0.05915769, 0.33655239])
```

The preceding output shows us that the most important features are the fourth and sixth ones, which correspond to **persons** and **safety**, respectively.

12. Import **ExtraTreesClassifier** from **sklearn.ensemble**:

```
from sklearn.ensemble import ExtraTreesClassifier
```

13. Instantiate **ExtraTreestClassifier** with **n_estimators=100**, **max_depth=6**, and **random_state=168**. Save it to a variable called **random_forest_classifier**:

```
extra_trees_classifier = \
ExtraTreesClassifier(n_estimators=100, \
                     max_depth=6, random_state=168)
```

14. Fit the **extratrees** classifier with the training set:

```
extra_trees_classifier.fit(features_train, label_train)
```

The output will be as follows:

```
ExtraTreesClassifier(bootstrap=False, class_weight=None, criterion='gini',
                     max_depth=6, max_features='auto', max_leaf_nodes=None,
                     min_impurity_decrease=0.0, min_impurity_split=None,
                     min_samples_leaf=1, min_samples_split=2,
                     min_weight_fraction_leaf=0.0, n_estimators=100,
                     n_jobs=None, oob_score=False, random_state=168, verbose=0,
                     warm_start=False)
```

Figure 4.20: Output with the extratrees classifier with the training set

These are the logs of the **extratrees** classifier with its hyperparameter values.

15. Make predictions on the testing set using the **extratrees** classifier and save them in a variable called **et_preds_test**. Print its content:

```
et_preds_test = extra_trees_classifier.predict(features_test)
et_preds_test
```

The output will be as follows:

```
array([0, 0, 2, 0, 2, 2, 3, 2, 2, 2, 2, 2, 2, 2, 2, 2, 2, 2, 2, 2, 2, 2,
       0, 2, 2, 0, 2, 2, 2, 2, 0, 2, 2, 2, 2, 2, 2, 2, 2, 2, 2, 2, 2, 2,
       2, 0, 2, 0, 0, 0, 2, 2, 2, 2, 0, 2, 0, 2, 2, 2, 0, 2, 2, 2, 2, 0,
       0, 2, 2, 2, 2, 2, 0, 2, 2, 2, 0, 2, 2, 2, 2, 0, 0, 0, 2, 2, 2,
       2, 2, 2, 2, 0, 2, 0, 0, 2, 2, 0, 2, 2, 2, 2, 0, 2, 0, 2, 2, 2, 2,
       0, 2, 2, 0, 2, 2, 0, 0, 2, 2, 2, 2, 0, 2, 2, 2, 2, 0, 2, 2, 2, 0,
       2, 2, 2, 2, 2, 2, 2, 0, 0, 0, 2, 0, 0, 2, 2, 2, 0, 0, 2, 2, 0, 2,
       2, 2, 2, 2, 0, 0, 2, 2, 2, 0, 2, 2, 2, 2, 2, 2, 0, 0, 2])
```

Figure 4.21: Predictions on the testing set using extratrees

16. Print the classification report with the labels and predictions from the test set:

```
print(classification_report(label_test, \
        extra_trees_classifier.predict(features_test)))
```

The output will be as follows:

```
              precision    recall  f1-score   support

           0       0.61      0.67      0.64        42
           1       0.00      0.00      0.00         9
           2       0.89      0.98      0.93       114
           3       1.00      0.12      0.22         8

    accuracy                           0.82       173
   macro avg       0.62      0.44      0.45       173
weighted avg       0.78      0.82      0.78       173
```

Figure 4.22: Classification report with the labels and predictions from the test set

The F_1 score shown in the preceding report shows us that the random forest is performing well on class **2** but not as good for class **0**. The model is unable to predict accurately for classes **1** and **3**, but there were only **9** and **8** observations in the testing set, respectively. The accuracy score is **0.82**, while the F_1 score is **0.78**. So, our **RandomForest** classifier performed better with **extratrees**.

17. Display the confusion matrix of the true and predicted labels of the testing set:

```
confusion_matrix(label_test, et_preds_test)
```

The output will be as follows:

```
array([[ 28,   0,  14,   0],
       [  9,   0,   0,   0],
       [  2,   0, 112,   0],
       [  7,   0,   0,   1]])
```

From this confusion matrix, we can see that the **extratrees** model is having difficulties accurately predicting the first and third classes.

18. Print the feature importance score on the test set using `.feature_importance_` and save the results in a variable called **et_varimp**. Print its content:

```
et_varimp = extra_trees_classifier.feature_importances_
et_varimp
```

The output will be as follows:

```
array([0.08844544, 0.0702334 , 0.01440408, 0.37662014, 0.05965896,
       0.39063797])
```

The preceding output shows us that the most important features are the sixth and fourth ones, which correspond to **safety** and **persons**, respectively. It is interesting to see that **RandomForest** has the same two most important features but in a different order.

> **NOTE**
>
> To access the source code for this specific section, please refer to https://packt.live/2YoUY5t.
>
> You can also run this example online at https://packt.live/3eswBcW.
>
> You must execute the entire Notebook in order to get the desired result.

CHAPTER 05: ARTIFICIAL INTELLIGENCE: CLUSTERING

ACTIVITY 5.01: CLUSTERING SALES DATA USING K-MEANS

Solution:

1. Open a new Jupyter Notebook file.

2. Load the dataset as a DataFrame and inspect the data:

```
import pandas as pd
file_url = 'https://raw.githubusercontent.com/'\
           'PacktWorkshops/'\
           'The-Applied-Artificial-Intelligence-Workshop/'\
           'master/Datasets/'\
           'Sales_Transactions_Dataset_Weekly.csv'
df = pd.read_csv(file_url)
df
```

The output of **df** is as follows:

	Product_Code	W0	W1	W2	W3	W4	W5	W6	W7	W8	...	Normalized 42	Normalized 43	Normalized 44	Normalized 45	Normalized 46	Normalized 47	Normalized 48
0	P1	11	12	10	8	13	12	14	21	6	...	0.06	0.22	0.28	0.39	0.50	0.00	0.22
1	P2	7	6	3	2	7	1	6	3	3	...	0.20	0.40	0.50	0.10	0.10	0.40	0.50
2	P3	7	11	8	9	10	8	7	13	12	...	0.27	1.00	0.18	0.18	0.36	0.45	1.00
3	P4	12	8	13	5	9	6	9	13	13	...	0.41	0.47	0.06	0.12	0.24	0.35	0.71
4	P5	8	5	13	11	6	7	9	14	9	...	0.27	0.53	0.27	0.60	0.20	0.20	0.13
...	...	...	...	...	...	...	...	...	...	...	...	...	...	...	...	...	...	...
806	P815	0	0	1	0	0	2	1	0	0	...	0.00	0.33	0.33	0.00	0.00	0.33	0.00
807	P816	0	1	0	0	1	2	2	6	0	...	0.43	0.43	0.57	0.29	0.57	0.71	0.71
808	P817	1	0	0	0	1	1	2	1	1	...	0.50	0.00	0.00	0.50	0.50	0.00	0.00
809	P818	0	0	0	1	0	0	0	0	1	...	0.00	0.00	0.00	0.50	0.50	0.00	0.00
810	P819	0	1	0	0	0	0	0	0	0	...	0.00	0.00	0.00	0.00	0.00	0.00	0.00

811 rows × 107 columns

Figure 5.18: Output showing the contents of the dataset

If you look at the output, you will notice that our dataset contains **811** rows, with each row representing a product. It also contains **107** columns, with the first column being the product code, then **52** columns starting with **W** representing the sale quantity for each week, and finally, the normalized version of the **52** columns, starting with the **Normalized** columns. The normalized columns will be a better choice to work with rather than the absolute sales columns, **W**, as they will help our k-means algorithms to find the center of each cluster faster. Since we are going to work on the normalized columns, we can remove every **W** column plus the **Product_Code** column. We can also remove the **MIN** and **MAX** columns as they do not bring any value to our clustering. Also notice that the weeks run from **0** to **51** and not **1** to **52**.

3. Next, create a new DataFrame without the unnecessary columns, as shown in the following code snippet (the first **55** columns of the dataset). You should use the **inplace** parameter to help you:

```
df2 = df.drop(df.iloc[:, 0:55], inplace = False, axis = 1)
```

The output of **df2** is as follows:

	Normalized 0	Normalized 1	Normalized 2	Normalized 3	Normalized 4	Normalized 5	Normalized 6	Normalized 7	Normalized 8	Normalized 9	Normalized 10
0	0.44	0.50	0.39	0.28	0.56	0.50	0.61	1.00	0.17	0.61	0.44
1	0.70	0.60	0.30	0.20	0.70	0.10	0.60	0.30	0.30	0.30	0.20
2	0.36	0.73	0.45	0.55	0.64	0.45	0.36	0.91	0.82	0.27	1.00
3	0.59	0.35	0.65	0.18	0.41	0.24	0.41	0.65	0.65	0.53	0.35
4	0.33	0.13	0.67	0.53	0.20	0.27	0.40	0.73	0.40	0.40	0.53
...	...	...	...	...	...	...	...	...	...	...	...
806	0.00	0.00	0.33	0.00	0.00	0.67	0.33	0.00	0.00	0.33	0.00
807	0.00	0.14	0.00	0.00	0.14	0.29	0.29	0.86	0.00	0.14	0.00
808	0.25	0.00	0.00	0.00	0.25	0.25	0.50	0.25	0.25	0.00	0.00
809	0.00	0.00	0.00	0.50	0.00	0.00	0.00	0.00	0.50	0.00	0.00
810	0.00	0.33	0.00	0.00	0.00	0.00	0.00	0.00	0.00	0.00	0.00

811 rows × 52 columns

Figure 5.19: Modified DataFrame

In the preceding code snippet, we used the **drop** function of the pandas DataFrame in order to remove the first **55** columns. We also set the **inplace** parameter to **False** in order to not remove the column of our original **df** DataFrame. As a result, we should only have the normalized columns from **0** to **51** in **df2** and **df** should still be unchanged.

4. Create a k-means clustering model with **8** clusters and with **random state = 8**:

```
from sklearn.cluster import KMeans
k_means_model = KMeans(n_clusters=8, random_state=8)
k_means_model.fit(df2)
```

We build a k-means model with the default value for every parameter except for **n_clusters=8** with **random_state=8** in order to obtain **8** clusters and reproducible results.

5. Retrieve the labels from the clustering algorithm:

```
labels = k_means_model.labels_
labels
```

The output of **labels** will be as follows:

```
array([4, 4, 3, 4, 4, 0, 4, 2, 2, 2, 4, 2, 3, 0, 3, 3, 3, 3, 3, 2, 3, 3,
       3, 3, 4, 0, 3, 3, 4, 3, 2, 3, 2, 4, 2, 4, 4, 4, 4, 3, 3, 3, 4, 3,
       4, 3, 4, 3, 3, 4, 3, 3, 0, 3, 4, 3, 3, 3, 3, 3, 3, 4, 3, 4, 2, 3,
       3, 4, 3, 3, 3, 3, 3, 4, 4, 3, 2, 3, 4, 4, 0, 4, 4, 4, 3, 3, 3, 3,
       3, 3, 2, 3, 2, 0, 0, 4, 3, 0, 4, 7, 3, 3, 3, 4, 4, 0, 2, 7, 2, 4,
       6, 4, 3, 3, 3, 0, 4, 4, 3, 3, 4, 4, 2, 0, 4, 4, 4, 3, 3, 3, 3, 4,
       3, 4, 3, 3, 3, 4, 3, 3, 3, 3, 3, 2, 4, 2, 3, 2, 3, 2, 4, 2, 4, 2,
       4, 2, 2, 4, 4, 2, 4, 4, 4, 2, 3, 4, 3, 4, 4, 3, 2, 3, 3, 3, 3, 3,
       4, 3, 3, 3, 3, 3, 4, 3, 4, 3, 3, 3, 4, 3, 3, 3, 4, 3, 4, 3, 3, 0,
       2, 0, 4, 0, 2, 2, 7, 0, 0, 0, 0, 0, 3, 1, 5, 1, 1, 5, 5, 5, 5, 5,
       6, 5, 1, 6, 5, 1, 1, 1, 1, 1, 5, 1, 1, 1, 5, 1, 1, 5, 1, 5, 5, 5,
       1, 5, 1, 1, 1, 5, 1, 5, 1, 5, 1, 1, 5, 1, 6, 1, 1, 1, 0, 0, 7, 2,
       7, 0, 2, 0, 0, 0, 7, 1, 1, 1, 1, 1, 5, 5, 1, 5, 7, 6, 1, 3, 0, 0,
       6, 6, 5, 5, 0, 7, 4, 4, 0, 0, 4, 2, 2, 2, 4, 2, 3, 2, 4, 4, 7, 4,
```

Figure 5.20: Output array of labels

It is very hard to make sense out of this output, but each index of **labels** represents the cluster that the product has been assigned, based on similar weekly sales trends. We can now use these cluster labels to group products together.

6. Now, from the first DataFrame, **df**, keep only the **W** columns and add the labels as a new column, as shown in the following code snippet:

```
df.drop(df.iloc[:, 53:], inplace = True, axis = 1)
df.drop('Product_Code', inplace = True, axis = 1)
df['label'] = labels
df
```

In the preceding code snippet, we removed all the unneeded columns and added **labels** as a new column in the DataFrame.

The output of **df** will be as follows:

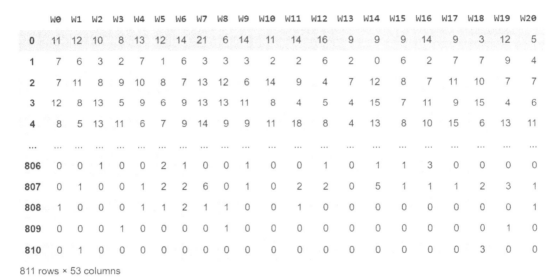

	W0	W1	W2	W3	W4	W5	W6	W7	W8	W9	W10	W11	W12	W13	W14	W15	W16	W17	W18	W19	W20
0	11	12	10	8	13	12	14	21	6	14	11	14	16	9	9	9	14	9	3	12	5
1	7	6	3	2	7	1	6	3	3	3	2	2	6	2	0	6	2	7	7	9	4
2	7	11	8	9	10	8	7	13	12	6	14	9	4	7	12	8	7	11	10	7	7
3	12	8	13	5	9	6	9	13	13	11	8	4	5	4	15	7	11	9	15	4	6
4	8	5	13	11	6	7	9	14	9	9	11	18	8	4	13	8	10	15	6	13	11
...	...	...	...	...	...	...	...	...	...	...	...	...	...	...	...	...	...	...	...	...	...
806	0	0	1	0	0	2	1	0	0	1	0	0	1	0	1	1	3	0	0	0	0
807	0	1	0	0	1	2	2	6	0	1	0	2	2	0	5	1	1	1	2	3	1
808	1	0	0	0	1	1	2	1	1	0	0	1	0	0	0	0	0	0	0	0	1
809	0	0	0	1	0	0	0	0	1	0	0	0	0	0	0	0	0	0	0	1	0
810	0	1	0	0	0	0	0	0	0	0	0	0	0	0	0	0	0	0	3	0	0

811 rows × 53 columns

Figure 5.21: Updated DataFrame with the new labels as a new column

Now that we have the label, we can perform aggregation on the **label** column in order to calculate the yearly average sales of each cluster.

7. Perform the aggregation (use the **groupby** function from pandas) in order to obtain the yearly average sale of each cluster, as shown in the following code snippet:

```
df_agg = df.groupby('label').sum()
df_final = df[['label','W0']].groupby('label').count()
df_final=df_final.rename(columns = {'W0':'count_product'})
df_final['total_sales'] = df_agg.sum(axis = 1)
df_final['yearly_average_sales']= \
```

```
df_final['total_sales'] / df_final['count_product']
df_final.sort_values(by='yearly_average_sales', \
                    ascending=False, inplace = True)
df_final
```

In the preceding code snippet, we first used the **groupby** function with the **sum()** method of the DataFrame to calculate the sum of every product's sales for each **W** column and cluster, and stored the results in **df_agg**. We then used the **groupby** function with the **count()** method on a single column (an arbitrary choice) of **df** to obtain the total number of products per cluster (note that we also had to rename the **W0** column after the aggregation). The next step was to sum all the sales columns of **df_agg** in order to obtain the total sales for each cluster. Finally, we calculated the **yearly_average_sales** for each cluster by dividing **total_sales** by **count_product**. We also included a final step to sort out the cluster by the highest **yearly_average_sales**.

The output of **df_final** will be as follows:

label	count_product	total_sales	yearly_average_sales
3	115	162668	1414.504348
4	113	86037	761.389381
0	109	57020	523.119266
2	87	39940	459.080460
7	88	21213	241.056818
6	89	6137	68.955056
5	87	1375	15.804598
1	123	897	7.292683

Figure 5.22: Expected output on the sales transaction dataset

Now, with this output, we see that our k-means model has managed to put similarly performing products together. We can easily see that the **115** products in cluster **3** are the best-selling products, whereas the **123** products of cluster **1** are performing very badly. This is very valuable for any business, as it helps them automatically identify and group together a number of similarly performing products without having any bias in the product name or description.

> **NOTE**
>
> To access the source code for this specific section, please refer to https://packt.live/3fVpSbT.
>
> You can also run this example online at https://packt.live/3hW24Gk.
>
> You must execute the entire Notebook in order to get the desired result.

By completing this activity, you have learned how to perform k-means clustering on multiple columns for many products. You have also learned how useful clustering can be for a business, even without label data.

ACTIVITY 5.02: CLUSTERING RED WINE DATA USING THE MEAN SHIFT ALGORITHM AND AGGLOMERATIVE HIERARCHICAL CLUSTERING

Solution:

1. Open a new Jupyter Notebook file.

2. Load the dataset as a DataFrame with **sep = ";"** and inspect the data:

```
import pandas as pd
import numpy as np
from sklearn import preprocessing
from sklearn.cluster import MeanShift
from sklearn.cluster import AgglomerativeClustering
from scipy.cluster.hierarchy import dendrogram
import scipy.cluster.hierarchy as sch
from sklearn import metrics
```

```
file_url = 'https://raw.githubusercontent.com/'\
           'PacktWorkshops/'\
           'The-Applied-Artificial-Intelligence-Workshop/'\
           'master/Datasets/winequality-red.csv'
df = pd.read_csv(file_url,sep=';')
df
```

The output of **df** is as follows:

	fixed acidity	volatile acidity	citric acid	residual sugar	chlorides	free sulfur dioxide	total sulfur dioxide	density	pH
0	7.4	0.700	0.00	1.9	0.076	11.0	34.0	0.99780	3.51
1	7.8	0.880	0.00	2.6	0.098	25.0	67.0	0.99680	3.20
2	7.8	0.760	0.04	2.3	0.092	15.0	54.0	0.99700	3.26
3	11.2	0.280	0.56	1.9	0.075	17.0	60.0	0.99800	3.16
4	7.4	0.700	0.00	1.9	0.076	11.0	34.0	0.99780	3.51
...	...	...	...	...	...	...	...	...	...
1594	6.2	0.600	0.08	2.0	0.090	32.0	44.0	0.99490	3.45
1595	5.9	0.550	0.10	2.2	0.062	39.0	51.0	0.99512	3.52
1596	6.3	0.510	0.13	2.3	0.076	29.0	40.0	0.99574	3.42
1597	5.9	0.645	0.12	2.0	0.075	32.0	44.0	0.99547	3.57
1598	6.0	0.310	0.47	3.6	0.067	18.0	42.0	0.99549	3.39

1599 rows × 12 columns

Figure 5.23: df showing the dataset as the output

NOTE

The output from the preceding screenshot is truncated.

Our dataset contains **1599** rows, with each row representing a red wine. It also contains **12** columns, with the last column being the quality of the wine. We can see that the remaining 11 columns will be our features, and we need to scale them in order to help the accuracy and speed of our models.

3. Create **features**, **label**, and **scaled_features** variables from the initial DataFrame, **df**:

```
features = df.drop('quality', 1)
label = df['quality']
scaled_features = preprocessing.scale(features)
```

In the preceding code snippet, we separated the label (**quality**) from the features. Then we used **preprocessing.scale** function from **sklearn** in order to scale our features, as this will improve our models.

4. Next, create a mean shift clustering model, then retrieve the model's predicted labels and the number of clusters created:

```
mean_shift_model = MeanShift()
mean_shift_model.fit(scaled_features)
n_cluster_mean_shift = len(mean_shift_model.cluster_centers_)
label_mean_shift = mean_shift_model.labels_
n_cluster_mean_shift
```

The output of **n_cluster_mean_shift** will be as follows:

```
10
```

Our mean shift model has created **10** clusters, which is already more than the number of groups that we have in our **quality** label. This will probably affect our extrinsic scores and might be an early indicator that wines sharing similar physicochemical properties don't belong in the same quality group.

The output of **label_mean_shift** will be as follows:

```
array([0, 0, 0, 0, 0, 0, 0, 0, 0, 0, 0, 0, 0, 3, 0, 0, 0, 1, 0, 1, 0, 0,
       0, 0, 0, 0, 0, 0, 0, 0, 0, 0, 0, 5, 0, 0, 0, 0, 0, 0, 0, 0, 1, 0,
       0, 0, 0, 0, 0, 0, 0, 0, 0, 0, 0, 0, 0, 0, 0, 0, 0, 0, 0, 0, 0, 0,
       0, 0, 0, 0, 0, 0, 0, 0, 0, 0, 0, 0, 0, 0, 0, 1, 0, 1, 0, 0, 3, 0,
       3, 0, 0, 3, 3, 0, 0, 0, 0, 0, 0, 0, 0, 0, 0, 0, 0, 0, 1, 0, 0, 0,
       0, 0, 0, 0, 0, 0, 0, 0, 0, 0, 0, 0, 0, 0, 0, 0, 0, 0, 0, 0, 0, 0,
       0, 0, 0, 0, 0, 0, 0, 0, 0, 0, 0, 0, 0, 0, 0, 0, 0, 0, 0, 8, 0, 0,
       0, 0, 0, 0, 0, 0, 0, 0, 0, 0, 0, 0, 0, 0, 0, 1, 0, 0, 0, 0, 0, 0,
       0, 0, 0, 0, 0, 1, 0, 0, 0, 0, 0, 0, 0, 0, 0, 0, 0, 0, 0, 0, 0, 0,
       0, 0, 0, 0, 0, 0, 0, 0, 0, 0, 0, 0, 0, 0, 0, 0, 0, 0, 0, 0, 0, 0,
       0, 0, 0, 0, 0, 0, 1, 0, 0, 0, 0, 0, 0, 0, 0, 0, 0, 0, 0, 0, 1, 0,
       0, 0, 0, 0, 0, 0, 0, 0, 0, 0, 0, 0, 0, 0, 0, 0, 1, 0, 0, 0, 0, 0,
       0, 0, 0, 0, 0, 0, 0, 0, 0, 0, 0, 0, 0, 0, 0, 0, 1, 0, 0, 0, 0, 0,
       0, 0, 0, 0, 0, 1, 0, 0, 0, 0, 0, 0, 0, 0, 0, 0, 0, 0, 0, 0, 0, 0,
       0, 0, 0, 0, 0, 0, 0, 0, 0, 0, 0, 0, 0, 0, 0, 0, 7, 7, 0, 0, 0, 0,
       0, 0, 0, 0, 0, 0, 0, 0, 0, 3, 0, 0, 0, 0, 0, 0, 0, 6, 0, 0, 0, 0,
       0, 6, 0, 0, 0, 0, 0, 0, 0, 0, 0, 0, 0, 0, 0, 0, 0, 0, 0, 0, 0, 0,
       6, 0, 0, 0, 6, 0, 0, 0, 0, 0, 0, 0, 0, 0, 0, 0, 0, 0, 0, 0, 0, 6,
       9, 0, 0, 0, 9, 0, 0, 0, 0, 0, 0, 0, 0, 0, 0, 0, 0, 0, 0, 0, 0, 0,
       0, 0, 0, 0, 0, 0, 0, 0, 0, 0, 0, 0, 0, 0, 0, 0, 0, 0, 0, 0, 0, 0,
       0, 0, 6, 0, 0, 0, 0, 0, 0, 0, 0, 1, 0, 0, 0, 6, 0, 0, 0, 0, 0, 0,
```

Figure 5.24: Output array of label_mean_shift

This is a very interesting output because it clearly shows that most wines in our dataset are very similar; there are a lot more wines in cluster **0** than in the other clusters.

5. Now create an agglomerative hierarchical clustering model after creating a dendrogram and selecting the optimal number of clusters for it:

```
dendrogram = sch.dendrogram(sch.linkage(scaled_features, \
                            method='ward'))
agglomerative_model = \
AgglomerativeClustering(n_clusters=7, \
                        affinity='euclidean', \
                        linkage='ward')
agglomerative_model.fit(scaled_features)
label_agglomerative = agglomerative_model.labels_
```

The output of **dendrogram** will be as follows:

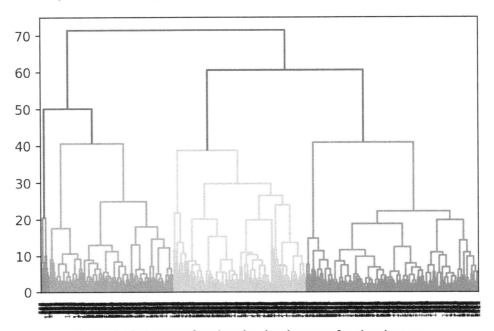

Figure 5.25: Output showing the dendrogram for the clusters

From this output, we can see that seven clusters seems to be the optimal number for our model. We get this number by searching for the highest difference on the *y* axis between the lowest branch and the highest branch. In our case, for seven clusters, the lowest branch has a value of **29** and the highest branch has a value of **41**.

The output of **label_agglomerative** will be as follows:

```
array([1, 1, 1, 0, 1, 1, 1, 1, 1, 6, 1, 6, 1, 0, 0, 0, 0, 3, 1, 3, 0, 0,
       2, 1, 0, 1, 1, 2, 1, 1, 1, 1, 0, 6, 0, 1, 1, 2, 1, 6, 6, 1, 3, 2,
       1, 4, 0, 2, 1, 0, 1, 1, 1, 0, 0, 1, 2, 6, 1, 1, 0, 0, 1, 1, 1, 1,
       1, 1, 0, 2, 1, 1, 1, 1, 0, 2, 2, 1, 1, 0, 1, 3, 0, 3, 2, 1, 0, 1,
       0, 1, 0, 0, 0, 1, 4, 4, 1, 1, 1, 1, 1, 1, 1, 1, 1, 1, 3, 1, 0, 0,
       2, 0, 0, 2, 2, 0, 1, 1, 1, 1, 1, 1, 1, 1, 0, 0, 1, 1, 2, 1, 0, 4,
       4, 1, 1, 1, 1, 0, 0, 0, 1, 1, 4, 1, 4, 0, 1, 3, 1, 1, 2, 3, 0, 0,
       6, 6, 6, 6, 1, 1, 1, 2, 1, 6, 6, 0, 1, 1, 1, 3, 1, 1, 1, 0, 1, 1,
       1, 1, 1, 1, 1, 3, 1, 1, 1, 0, 0, 1, 0, 0, 0, 0, 0, 1, 1, 0, 1, 5,
       4, 1, 2, 0, 1, 0, 0, 5, 5, 0, 0, 2, 2, 1, 5, 0, 1, 6, 1, 1, 1, 0,
       0, 1, 1, 1, 1, 0, 3, 1, 0, 1, 4, 1, 0, 1, 1, 1, 1, 1, 1, 1, 3, 5,
       0, 5, 5, 1, 1, 1, 1, 1, 2, 1, 2, 0, 1, 0, 2, 1, 3, 2, 0, 1, 1, 0,
       5, 5, 1, 2, 1, 2, 0, 2, 0, 1, 6, 0, 1, 2, 2, 6, 5, 3, 1, 6, 0, 0,
       0, 0, 0, 0, 0, 3, 0, 1, 5, 5, 0, 1, 1, 1, 1, 5, 1, 1, 0, 5, 1, 3,
       3, 1, 5, 0, 0, 0, 0, 0, 0, 0, 0, 0, 0, 1, 0, 6, 6, 5, 2, 5, 2,
       2, 2, 0, 1, 0, 5, 2, 0, 0, 5, 5, 2, 2, 2, 5, 0, 4, 5, 0, 1, 1, 1,
       1, 5, 0, 4, 0, 2, 5, 5, 1, 0, 5, 5, 5, 2, 5, 5, 0, 2, 0, 2, 0, 1,
       5, 5, 2, 2, 5, 0, 2, 5, 2, 2, 0, 1, 0, 1, 0, 0, 4, 5, 2, 0, 5, 5,
       6, 5, 5, 1, 6, 2, 5, 5, 1, 2, 2, 5, 2, 0, 0, 0, 1, 2, 0, 6, 2, 0,
       5, 0, 0, 4, 1, 2, 1, 4, 4, 1, 2, 5, 2, 1, 2, 5, 0, 5, 0, 5, 0, 1,
       5, 5, 5, 2, 4, 1, 0, 0, 1, 5, 5, 3, 1, 5, 2, 5, 1, 0, 5, 5, 2, 1,
       2, 0, 5, 2, 0, 2, 5, 0, 5, 0, 0, 2, 2, 1, 2, 2, 1, 1, 6, 2, 5, 5,
```

Figure 5.26: Array showing label_agglomerative

We can see that we have a predominant cluster, **1**, but not as much as was the case in the mean shift model.

6. Now, compute the following extrinsic approach scores for both models:

a. Begin with the adjusted Rand index:

```
ARI_mean=metrics.adjusted_rand_score(label, label_mean_shift)
ARI_agg=metrics.adjusted_rand_score(label, label_agglomerative)
ARI_mean
```

The output of **ARI_mean** will be as follows:

```
0.0006771608724007207
```

Next, enter **ARI_agg** to get the expected values:

```
ARI_agg
```

The output of **ARI_agg** will be as follows:

```
0.05358047852603172
```

Our agglomerative model has a much higher **adjusted_rand_score** than the mean shift model, but both scores are very close to **0**, which means that neither model is performing very well with regard to the true labels.

b. Next, calculate the adjusted mutual information:

```
AMI_mean = metrics.adjusted_mutual_info_score(label, \
                                    label_mean_shift)
AMI_agg = metrics.adjusted_mutual_info_score(label, \
                                    label_agglomerative)

AMI_mean
```

The output of **AMI_mean** will be as follows:

```
0.004837187596124968
```

Next, enter **AMI_agg** to get the expected values:

```
AMI_agg
```

The output of **AMI_agg** will be as follows:

```
0.05993098663692826
```

Our agglomerative model has a much higher **adjusted_mutual_info_score** than the mean shift model, but both scores are very close to **0**, which means that neither model is performing very well with regard to the true labels.

c. Calculate the V-Measure:

```
V_mean = metrics.v_measure_score(label, \
                            label_mean_shift, beta=1)
V_agg = metrics.v_measure_score(label, \
                            label_agglomerative, beta=1)

V_mean
```

The output of **V_mean** will be as follows:

```
0.021907254751144124
```

Next, enter **V_agg** to get the expected values:

```
V_agg
```

The output of **V_agg** will be as follows:

```
0.07549735446050691
```

Our agglomerative model has a higher V-Measure than the mean shift model, but both scores are very close to **0**, which means that neither model is performing very well with regard to the true labels.

d. Next, find the Fowlkes-Mallows score:

```
FM_mean = metrics.fowlkes_mallows_score(label, \
                                        label_mean_shift)
FM_agg=  metrics.fowlkes_mallows_score(label, \
                                        label_agglomerative)
FM_mean
```

The output of **FM_mean** will be as follows:

```
0.5721233634622408
```

Next, enter **FM_agg** to get the expected values:

```
FM_agg
```

The output of **FM_agg** will be as follows:

```
0.3300681478007641
```

This time, our mean shift model has a higher Fowlkes-Mallows score than the agglomerative model, but both scores are still on the lower range of the score, which means that neither model is performing very well with regard to the true labels.

In conclusion, with the extrinsic approach evaluation, neither of our models were able to find clusters containing wines of a similar quality based on their physicochemical properties. We will confirm this by using the intrinsic approach evaluation to ensure that our models' clusters are well defined and are properly grouping similar wines together.

7. Now, compute the following intrinsic approach scores for both models:

    a. Begin with the Silhouette Coefficient:

    ```
    Sil_mean = metrics.silhouette_score(scaled_features, \
                                     label_mean_shift)
    Sil_agg = metrics.silhouette_score(scaled_features, \
                                    label_agglomerative)
    Sil_mean
    ```

    The output of **Sil_mean** will be as follows:

    ```
    0.32769323700400077
    ```

    Next, enter **Sil_agg** to get the expected values:

    ```
    Sil_agg
    ```

    The output of **Sil_agg** will be as follows:

    ```
    0.1591882574407987
    ```

    Our mean shift model has a higher Silhouette Coefficient than the agglomerative model, but both scores are very close to **0**, which means that both models have overlapping clusters.

    b. Next, find the Calinski-Harabasz index:

    ```
    CH_mean = metrics.calinski_harabasz_score(scaled_features, \
                                      label_mean_shift)
    CH_agg = metrics.calinski_harabasz_score(scaled_features, \
                                     label_agglomerative)
    CH_mean
    ```

    The output of **CH_mean** will be as follows:

    ```
    44.62091774102674
    ```

    Next, enter **CH_agg** to get the expected values:

    ```
    CH_agg
    ```

    The output of **CH_agg** will be as follows:

    ```
    223.5171774491095
    ```

Our agglomerative model has a much higher Calinski-Harabasz index than the mean shift model, which means that the agglomerative model has much more dense and well-defined clusters than the mean shift model.

c. Finally, find the Davies-Bouldin index:

```
DB_mean = metrics.davies_bouldin_score(scaled_features, \
                                       label_mean_shift)
DB_agg = metrics.davies_bouldin_score(scaled_features, \
                                      label_agglomerative)
DB_mean
```

The output of **DB_mean** will be as follows:

```
0.8106334674570222
```

Next, enter **DB_agg** to get the expected values:

```
DB_agg
```

The output of **DB_agg** will be as follows:

```
1.4975443816135114
```

Our agglomerative model has a higher David-Bouldin index than the mean shift model, but both scores are close to **0**, which means that both models are performing well with regard to the definition of their clusters.

> **NOTE**
>
> To access the source code for this specific section, please refer to https://packt.live/2YXMI0U.
>
> You can also run this example online at https://packt.live/2Bs7sAp.
>
> You must execute the entire Notebook in order to get the desired result.

In conclusion, with the intrinsic approach evaluation, both our models were well defined and confirm our intuition on the red wine dataset, that is, similar physicochemical properties are not associated with similar quality. We were also able to see that in most of our scores, the agglomerative hierarchical model performs better than the mean shift model.

CHAPTER 06: NEURAL NETWORKS AND DEEP LEARNING

ACTIVITY 6.01: FINDING THE BEST ACCURACY SCORE FOR THE DIGITS DATASET

Solution:

1. Open a new Jupyter Notebook file.

2. Import **tensorflow.keras.datasets.mnist** as **mnist**:

   ```
   import tensorflow.keras.datasets.mnist as mnist
   ```

3. Load the **mnist** dataset using **mnist.load_data()** and save the results into **(features_train, label_train), (features_test, label_test)**:

   ```
   (features_train, label_train), \
   (features_test, label_test) = mnist.load_data()
   ```

4. Print the content of **label_train**:

   ```
   label_train
   ```

   The expected output is this:

   ```
   array([5, 0, 4, ..., 5, 6, 8], dtype=uint8)
   ```

   The **label** column contains numeric values that correspond to the **10** handwritten digits: **0** to **9**.

5. Print the shape of the training set:

   ```
   features_train.shape
   ```

   The expected output is this:

   ```
   (60000, 28, 28)
   ```

   The training set is composed of **60,000** observations of shape **28** by **28**. We will need to flatten the input for our neural network.

6. Print the shape of the testing set:

   ```
   features_test.shape
   ```

   The expected output is this:

   ```
   (10000, 28, 28)
   ```

   The testing set is composed of **10,000** observations of shape **28** by **28**.

7. Standardize **features_train** and **features_test** by dividing them by **255**:

```
features_train = features_train / 255.0
features_test = features_test / 255.0
```

8. Import **numpy** as **np**, **tensorflow** as **tf**, and **layers** from **tensorflow.keras**:

```
import numpy as np
import tensorflow as tf
from tensorflow.keras import layers
```

9. Set **8** as the seed for NumPy and TensorFlow using **np.random_seed()** and **tf.random.set_seed()**:

```
np.random.seed(8)
tf.random.set_seed(8)
```

10. Instantiate a **tf.keras.Sequential()** class and save it into a variable called **model**:

```
model = tf.keras.Sequential()
```

11. Instantiate **layers.Flatten()** with **input_shape=(28,28)** and save it into a variable called **input_layer**:

```
input_layer = layers.Flatten(input_shape=(28,28))
```

12. Instantiate a **layers.Dense()** class with **128** neurons and **activation='relu'**, then save it into a variable called **layer1**:

```
layer1 = layers.Dense(128, activation='relu')
```

13. Instantiate a second **layers.Dense()** class with **1** neuron and **activation='softmax'**, then save it into a variable called **final_layer**:

```
final_layer = layers.Dense(10, activation='softmax')
```

14. Add the three layers you just defined to the model using **.add()** and add a **layers.Dropout(0.25)** layer in between each of them (except for the flatten layer):

```
model.add(input_layer)
model.add(layer1)
model.add(layers.Dropout(0.25))
model.add(final_layer)
```

15. Instantiate a **tf.keras.optimizers.Adam()** class with **0.001** as learning rate and save it into a variable called **optimizer**:

```
optimizer = tf.keras.optimizers.Adam(0.001)
```

16. Compile the neural network using .**compile()** with **loss='sparse_categorical_crossentropy', optimizer=optimizer, metrics=['accuracy']**:

```
model.compile(loss='sparse_categorical_crossentropy', \
              optimizer=optimizer, \
              metrics=['accuracy'])
```

17. Print a summary of the model using .**summary()**:

```
model.summary()
```

The expected output is this:

```
Model: "sequential"
```

Layer (type)	Output Shape	Param #
flatten (Flatten)	(None, 784)	0
dense (Dense)	(None, 128)	100480
dropout (Dropout)	(None, 128)	0
dense_1 (Dense)	(None, 10)	1290

```
Total params: 101,770
Trainable params: 101,770
Non-trainable params: 0
```

Figure 6.29: Summary of the model

This output summarizes the architecture of our neural networks. We can see it is composed of four layers with one flatten layer, two dense layers, and one dropout layer.

18. Instantiate the **tf.keras.callbacks.EarlyStopping()** class with **monitor='val_loss'** and **patience=5** as the learning rate and save it into a variable called **callback**:

```
callback = tf.keras.callbacks.EarlyStopping(monitor='val_loss', \
                                            patience=5)
```

19. Fit the neural networks with the training set and specify **epochs=10**, **validation_split=0.2**, **callbacks=[callback]**, and **verbose=2**:

```
model.fit(features_train, label_train, epochs=10, \
          validation_split = 0.2, \
          callbacks=[callback], verbose=2)
```

The expected output is this:

```
Train on 48000 samples, validate on 12000 samples
Epoch 1/10
48000/48000 - 3s - loss: 0.3383 - accuracy: 0.9007 - val_loss: 0.1580 - val_accuracy: 0.9540
Epoch 2/10
48000/48000 - 2s - loss: 0.1666 - accuracy: 0.9509 - val_loss: 0.1235 - val_accuracy: 0.9645
Epoch 3/10
48000/48000 - 2s - loss: 0.1274 - accuracy: 0.9612 - val_loss: 0.1043 - val_accuracy: 0.9706
Epoch 4/10
48000/48000 - 2s - loss: 0.1036 - accuracy: 0.9688 - val_loss: 0.0973 - val_accuracy: 0.9705
Epoch 5/10
48000/48000 - 2s - loss: 0.0877 - accuracy: 0.9731 - val_loss: 0.0832 - val_accuracy: 0.9748
Epoch 6/10
48000/48000 - 2s - loss: 0.0774 - accuracy: 0.9759 - val_loss: 0.0838 - val_accuracy: 0.9752
Epoch 7/10
48000/48000 - 2s - loss: 0.0679 - accuracy: 0.9781 - val_loss: 0.0830 - val_accuracy: 0.9749
Epoch 8/10
48000/48000 - 2s - loss: 0.0637 - accuracy: 0.9795 - val_loss: 0.0813 - val_accuracy: 0.9768
Epoch 9/10
48000/48000 - 2s - loss: 0.0576 - accuracy: 0.9809 - val_loss: 0.0851 - val_accuracy: 0.9766
Epoch 10/10
48000/48000 - 2s - loss: 0.0523 - accuracy: 0.9825 - val_loss: 0.0787 - val_accuracy: 0.9779

<tensorflow.python.keras.callbacks.History at 0x161c8ed50>
```

Figure 6.30: Fitting the neural network with the training set

We achieved an accuracy score of **0.9825** for the training set and **0.9779** for the validation set for recognizing hand-written digits after just **10** epochs. These are amazing results. In this section, you learned how to build and train a neural network from scratch using TensorFlow to classify digits.

> **NOTE**
>
> To access the source code for this specific section, please refer to https://packt.live/37UWf7E.
>
> You can also run this example online at https://packt.live/317R2b3.
>
> You must execute the entire Notebook in order to get the desired result.

ACTIVITY 6.02: EVALUATING A FASHION IMAGE RECOGNITION MODEL USING CNNS

Solution:

1. Open a new Jupyter Notebook.

2. Import **tensorflow.keras.datasets.fashion_mnist** as **fashion_mnist**:

    ```
    import tensorflow.keras.datasets.fashion_mnist as fashion_mnist
    ```

3. Load the Fashion MNIST dataset using **fashion_mnist.load_data()** and save the results into **(features_train, label_train)**, **(features_test, label_test)**:

    ```
    (features_train, label_train), \
    (features_test, label_test) = fashion_mnist.load_data()
    ```

4. Print the shape of the training set:

    ```
    features_train.shape
    ```

    The expected output is this:

    ```
    (60000, 28, 28)
    ```

    The training set is composed of **60,000** images of size **28*28**.

5. Print the shape of the testing set:

```
features_test.shape
```

The expected output is this:

```
(10000, 28, 28)
```

The testing set is composed of **10,000** images of size **28*28**.

6. Reshape the training and testing sets with the dimensions (**number_rows**, **28, 28, 1**), as shown in the following code snippet:

```
features_train = features_train.reshape(60000, 28, 28, 1)
features_test = features_test.reshape(10000, 28, 28, 1)
```

7. Standardize **features_train** and **features_test** by dividing them by **255**:

```
features_train = features_train / 255.0
features_test = features_test / 255.0
```

8. Import **numpy** as **np**, **tensorflow** as **tf**, and **layers** from **tensorflow.keras**:

```
import numpy as np
import tensorflow as tf
from tensorflow.keras import layers
```

9. Set **8** as the seed for **numpy** and **tensorflow** using **np.random_seed()** and **tf.random.set_seed()**:

```
np.random.seed(8)
tf.random.set_seed(8)
```

10. Instantiate a **tf.keras.Sequential()** class and save it into a variable called **model**:

```
model = tf.keras.Sequential()
```

11. Instantiate **layers.Conv2D()** with **64** kernels of shape **(3,3)**, **activation='relu' and input_shape=(28,28)** and save it into a variable called **conv_layer1**:

```
conv_layer1 = layers.Conv2D(64, (3,3), \
              activation='relu', input_shape=(28, 28, 1))
```

12. Instantiate **layers.Conv2D()** with **64** kernels of shape **(3,3)**, **activation='relu'** and save it into a variable called **conv_layer2**:

```
conv_layer2 = layers.Conv2D(64, (3,3), activation='relu')
```

13. Instantiate **layers.Flatten()** with **128** neurons and **activation='relu'**, then save it into a variable called **fc_layer1**:

```
fc_layer1 = layers.Dense(128, activation='relu')
```

14. Instantiate **layers.Flatten()** with **10** neurons and **activation='softmax'**, then save it into a variable called **fc_layer2**:

```
fc_layer2 = layers.Dense(10, activation='softmax')
```

15. Add the four layers you just defined to the model using **.add()** and add a **MaxPooling2D()** layer of size **(2,2)** in between each of the convolutional layers:

```
model.add(conv_layer1)
model.add(layers.MaxPooling2D(2, 2))
model.add(conv_layer2)
model.add(layers.MaxPooling2D(2, 2))
model.add(layers.Flatten())
model.add(fc_layer1)
model.add(fc_layer2)
```

16. Instantiate a **tf.keras.optimizers.Adam()** class with **0.001** as the learning rate and save it into a variable called **optimizer**:

```
optimizer = tf.keras.optimizers.Adam(0.001)
```

17. Compile the neural network using **.compile()** with **loss='sparse_ categorical_crossentropy', optimizer=optimizer, metrics=['accuracy']**:

```
model.compile(loss='sparse_categorical_crossentropy', \
              optimizer=optimizer, metrics=['accuracy'])
```

18. Print a summary of the model using **.summary()**:

```
model.summary()
```

The expected output is this:

```
Model: "sequential"
```

Layer (type)	Output Shape	Param #
conv2d (Conv2D)	(None, 26, 26, 64)	640
max_pooling2d (MaxPooling2D)	(None, 13, 13, 64)	0
conv2d_1 (Conv2D)	(None, 11, 11, 64)	36928
max_pooling2d_1 (MaxPooling2	(None, 5, 5, 64)	0
flatten (Flatten)	(None, 1600)	0
dense (Dense)	(None, 128)	204928
dense_1 (Dense)	(None, 10)	1290

```
Total params: 243,786
Trainable params: 243,786
Non-trainable params: 0
```

Figure 6.31: Summary of the model

The summary shows us that there are more than **240,000** parameters to be optimized with this model.

19. Fit the neural network with the training set and specify **epochs=5**, **validation_split=0.2**, and **verbose=2**:

```
model.fit(features_train, label_train, \
          epochs=5, validation_split = 0.2, verbose=2)
```

The expected output is this:

```
Train on 48000 samples, validate on 12000 samples
Epoch 1/5
48000/48000 - 32s - loss: 0.4652 - accuracy: 0.8309 - val_loss: 0.3393 - val_accuracy: 0.8761
Epoch 2/5
48000/48000 - 33s - loss: 0.3107 - accuracy: 0.8861 - val_loss: 0.2958 - val_accuracy: 0.8947
Epoch 3/5
48000/48000 - 34s - loss: 0.2628 - accuracy: 0.9023 - val_loss: 0.2752 - val_accuracy: 0.8979
Epoch 4/5
48000/48000 - 33s - loss: 0.2295 - accuracy: 0.9145 - val_loss: 0.2673 - val_accuracy: 0.9019
Epoch 5/5
48000/48000 - 35s - loss: 0.2025 - accuracy: 0.9250 - val_loss: 0.2661 - val_accuracy: 0.9042

<tensorflow.python.keras.callbacks.History at 0x16939b910>
```

Figure 6.32: Fitting the neural network with the training set

After training for **5** epochs, we achieved an accuracy score of **0.925** for the training set and **0.9042** for the validation set. Our model is overfitting a bit.

20. Evaluate the performance of the model on the testing set:

```
model.evaluate(features_test, label_test)
```

The expected output is this:

```
10000/10000 [==============================] - 1s 108us/sample -
loss: 0.2746 - accuracy: 0.8976
[0.27461639745235444, 0.8976]
```

We achieved an accuracy score of **0.8976** on the testing set for predicting images of clothing from the Fashion MNIST dataset. You can try on your own to improve this score and reduce the overfitting.

> **NOTE**
>
> To access the source code for this specific section, please refer to https://packt.live/2Nzt6pn.
>
> You can also run this example online at https://packt.live/2NlM5nd.
>
> You must execute the entire Notebook in order to get the desired result.

In this activity, we designed and trained a CNN architecture for recognizing images of clothing from the Fashion MNIST dataset.

ACTIVITY 6.03: EVALUATING A YAHOO STOCK MODEL WITH AN RNN

Solution:

1. Open a Jupyter Notebook.

2. Import **pandas** as **pd** and **numpy** as **np**:

    ```
    import pandas as pd
    import numpy as np
    ```

3. Create a variable called **file_url** containing a link to the raw dataset:

    ```
    file_url = 'https://raw.githubusercontent.com/'\
                'PacktWorkshops/'\
                'The-Applied-Artificial-Intelligence-Workshop/'\
                'master/Datasets/yahoo_spx.csv'
    ```

4. Load the dataset using **pd.read_csv()** into a new variable called **df**:

    ```
    df = pd.read_csv(file_url)
    ```

5. Extract the values of the second column using **.iloc** and **.values** and save the results in a variable called **stock_data**:

    ```
    stock_data = df.iloc[:, 1:2].values
    ```

6. Import **MinMaxScaler** from **sklearn.preprocessing**:

    ```
    from sklearn.preprocessing import MinMaxScaler
    ```

7. Instantiate **MinMaxScaler()** and save it to a variable called **sc**:

    ```
    sc = MinMaxScaler()
    ```

8. Standardize the data with **.fit_transform()** and save the results in a variable called **stock_data_scaled**:

    ```
    stock_data_scaled = sc.fit_transform(stock_data)
    ```

9. Create two empty arrays called **X_data** and **y_data**:

    ```
    X_data = []
    y_data = []
    ```

10. Create a variable called **window** that will contain the value **30**:

    ```
    window = 30
    ```

11. Create a **for** loop starting from the **window** value and iterate through the length of the dataset. For each iteration, append to **X_data** the previous rows of **stock_data_scaled** using **window** and append the current value of **stock_data_scaled**:

```
for i in range(window, len(df)):
    X_data.append(stock_data_scaled[i - window:i, 0])
    y_data.append(stock_data_scaled[i, 0])
```

y_data will contain the opening stock price for each day and **X_data** will contain the last 30 days' stock prices.

12. Convert **X_data** and **y_data** into NumPy arrays:

```
X_data = np.array(X_data)
y_data = np.array(y_data)
```

13. Reshape **X_data** as (number of rows, number of columns, 1):

```
X_data = np.reshape(X_data, (X_data.shape[0], \
                    X_data.shape[1], 1))
```

14. Use the first **1,000** rows as the training data and save them into two variables called **features_train** and **label_train**:

```
features_train = X_data[:1000]
label_train = y_data[:1000]
```

15. Use the rows after row **1,000** as the testing data and save them into two variables called **features_test** and **label_test**:

```
features_test = X_data[:1000]
label_test = y_data[:1000]
```

16. Import **numpy** as **np**, **tensorflow** as **tf**, and **layers** from **tensorflow.keras**:

```
import numpy as np
import tensorflow as tf
from tensorflow.keras import layers
```

17. Set **8** as **seed** for NumPy and TensorFlow using **np.random_seed()** and **tf.random.set_seed()**:

```
np.random.seed(8)
tf.random.set_seed(8)
```

18. Instantiate a **tf.keras.Sequential()** class and save it into a variable called **model**:

```
model = tf.keras.Sequential()
```

19. Instantiate **layers.LSTM()** with **50** units, **return_sequences='True'**, and **input_shape=(X_train.shape[1], 1)**, then save it into a variable called **lstm_layer1**:

```
lstm_layer1 = layers.LSTM(units=50,return_sequences=True,\
                      input_shape=(features_train.shape[1], 1))
```

20. Instantiate **layers.LSTM()** with **50** units and **return_sequences='True'**, then save it into a variable called **lstm_layer2**:

```
lstm_layer2 = layers.LSTM(units=50,return_sequences=True)
```

21. Instantiate **layers.LSTM()** with **50** units and **return_sequences='True'**, then save it into a variable called **lstm_layer3**:

```
lstm_layer3 = layers.LSTM(units=50,return_sequences=True)
```

22. Instantiate **layers.LSTM()** with **50** units and save it into a variable called **lstm_layer4**:

```
lstm_layer4 = layers.LSTM(units=50)
```

23. Instantiate **layers.Dense()** with **1** neuron and save it into a variable called **fc_layer**:

```
fc_layer = layers.Dense(1)
```

24. Add the five layers you just defined to the model using **.add()** and add a **Dropout(0.2)** layer in between each of the LSTM layers:

```
model.add(lstm_layer1)
model.add(layers.Dropout(0.2))
model.add(lstm_layer2)
model.add(layers.Dropout(0.2))
model.add(lstm_layer3)
model.add(layers.Dropout(0.2))
model.add(lstm_layer4)
model.add(layers.Dropout(0.2))
model.add(fc_layer)
```

25. Instantiate a **tf.keras.optimizers.Adam()** class with **0.001** as the learning rate and save it into a variable called **optimizer**:

```
optimizer = tf.keras.optimizers.Adam(0.001)
```

26. Compile the neural network using **.compile()** with **loss='mean_squared_error', optimizer=optimizer, metrics=[mse]**:

```
model.compile(loss='mean_squared_error', \
              optimizer=optimizer, metrics=['mse'])
```

27. Print a summary of the model using **.summary()**:

```
model.summary()
```

The expected output is this:

```
Model: "sequential_3"
```

Layer (type)	Output Shape	Param #
lstm_12 (LSTM)	(None, 30, 50)	10400
dropout_10 (Dropout)	(None, 30, 50)	0
lstm_13 (LSTM)	(None, 30, 50)	20200
dropout_11 (Dropout)	(None, 30, 50)	0
lstm_14 (LSTM)	(None, 30, 50)	20200
dropout_12 (Dropout)	(None, 30, 50)	0
lstm_15 (LSTM)	(None, 50)	20200
dropout_13 (Dropout)	(None, 50)	0
dense_3 (Dense)	(None, 1)	51

```
Total params: 71,051
Trainable params: 71,051
Non-trainable params: 0
```

Figure 6.33: Summary of the model

The summary shows us that there are more than **71,051** parameters to be optimized with this model.

28. Fit the neural network with the training set and specify **epochs=10, validation_split=0.2, verbose=2**:

```
model.fit(features_train, label_train, epochs=10, \
          validation_split = 0.2, verbose=2)
```

The expected output is this:

```
Train on 800 samples, validate on 200 samples
Epoch 1/10
800/800 - 6s - loss: 0.0235 - mse: 0.0235 - val_loss: 0.0056 - val_mse: 0.0056
Epoch 2/10
800/800 - 1s - loss: 0.0038 - mse: 0.0038 - val_loss: 0.0063 - val_mse: 0.0063
Epoch 3/10
800/800 - 1s - loss: 0.0028 - mse: 0.0028 - val_loss: 0.0035 - val_mse: 0.0035
Epoch 4/10
800/800 - 1s - loss: 0.0029 - mse: 0.0029 - val_loss: 0.0044 - val_mse: 0.0044
Epoch 5/10
800/800 - 1s - loss: 0.0027 - mse: 0.0027 - val_loss: 0.0054 - val_mse: 0.0054
Epoch 6/10
800/800 - 1s - loss: 0.0034 - mse: 0.0034 - val_loss: 0.0034 - val_mse: 0.0034
Epoch 7/10
800/800 - 1s - loss: 0.0029 - mse: 0.0029 - val_loss: 0.0046 - val_mse: 0.0046
Epoch 8/10
800/800 - 1s - loss: 0.0030 - mse: 0.0030 - val_loss: 0.0039 - val_mse: 0.0039
Epoch 9/10
800/800 - 1s - loss: 0.0025 - mse: 0.0025 - val_loss: 0.0035 - val_mse: 0.0035
Epoch 10/10
800/800 - 1s - loss: 0.0025 - mse: 0.0025 - val_loss: 0.0033 - val_mse: 0.0033

<tensorflow.python.keras.callbacks.History at 0x146157950>
```

Figure 6.34: Fitting the neural network with the training set

After training for **10** epochs, we achieved a mean squared error score of **0.0025** for the training set and **0.0033** for the validation set. Our model is overfitting a little bit.

29. Finally, evaluate the performance of the model on the testing set:

```
model.evaluate(features_test, label_test)
```

The expected output is this:

```
1000/1000 [==============================] - 0s 279us/sample - loss: 0.0016 - mse: 0.0016
[0.00158528157370165, 0.0015852816]
```

We achieved a mean squared error score of **0.0017** on the testing set, which means we can quite accurately predict the stock price of Yahoo using the last 30 days' stock price data as features.

> **NOTE**
>
> To access the source code for this specific section, please refer to https://packt.live/3804U8P.
>
> You can also run this example online at https://packt.live/3hWtU5I.
>
> You must execute the entire Notebook in order to get the desired result.

In this activity, we designed and trained an RNN model to predict the Yahoo stock price from the previous 30 days of data.

INDEX

www.ingramcontent.com/pod-product-compliance
Lightning Source LLC
Chambersburg PA
CBHW081503050326
40690CB00015B/2907